Lecture Notes Editorial Policies

Lecture Notes in Statistics provides a format for the informal and quick publication of monographs, case studies, and workshops of theoretical or applied importance. Thus, in some instances, proofs may be merely outlined and results presented which will later be published in a different form.

Publication of the Lecture Notes is intended as a service to the international statistical community, in that a commercial publisher, Springer-Verlag, can provide efficient distribution of documents that would otherwise have a restricted readership. Once published and copyrighted, they can be documented and discussed in the scientific literature.

Lecture Notes are reprinted photographically from the copy delivered in camera-ready form by the author or editor. Springer-Verlag provides technical instructions for the preparation of manuscripts. Volumes should be no less than 100 pages and preferably no more than 400 pages. A subject index is expected for authored but not edited volumes. Proposals for volumes should be sent to one of the series editors or addressed to "Statistics Editor" at Springer-Verlag in New York.

Authors of monographs receive 50 free copies of their book. Editors receive 50 free copies and are responsible for distributing them to contributors. Authors, editors, and contributors may purchase additional copies at the publisher's discount. No reprints of individual contributions will be supplied and no royalties are paid on Lecture Notes volumes. Springer-Verlag secures the copyright for each volume.

Series Editors:

Professor P. Bickel
Department of Statistics
University of California
Berkeley, California 94720
USA

Professor P. Diggle
Department of Mathematics
Lancaster University
Lancaster LA1 4YL
England

Professor S. Fienberg
Department of Statistics
Carnegie Mellon University
Pittsburgh, Pennsylvania 15213
USA

Professor K. Krickeberg
3 Rue de L'Estrapade
75005 Paris
France

Professor I. Olkin
Department of Statistics
Stanford University
Stanford, California 94305
USA

Professor N. Wermuth
Department of Psychology
Johannes Gutenberg University
Postfach 3980
D-6500 Mainz
Germany

Professor S. Zeger
Department of Biostatistics
The Johns Hopkins University
615 N. Wolfe Street
Baltimore, Maryland 21205-2103
USA

Lecture Notes in Statistics 154

Edited by P. Bickel, P. Diggle, S. Fienberg, K. Krickeberg, I. Olkin, N. Wermuth, S. Zeger

Springer Science+Business Media, LLC

Regina Kaiser
Agustín Maravall

Measuring Business Cycles in Economic Time Series

 Springer

Regina Kaiser
D. de Estadística y Econometría
Universidad Carlos III de Madrid
Madrid, 126
28903 Getafe
Spain
kaiser@est-econ.uc3m.es

Agustín Maravall
Servicio de Estudios
Banco de España
Alcalá 50
28014 Madrid
Spain
maravall@bde.es

The first author acknowledges support by the Spanish grant PB95-0299 of CICYT.

Library of Congress Cataloging-in-Publication Data

Kaiser, Regina.
    Measuring business cycles in economic time series / Regina Kaiser, Agustín Maravall.
       p.   cm.—(Lecture notes in statistics; 154)
    Includes bibliographical references and indexes.
    ISBN 978-0-387-95112-6   ISBN 978-1-4613-0129-5 (eBook)
    DOI 10.1007/978-1-4613-0129-5
    1. Business cycles. 2. Time-series analysis. I. Maravall, Agustín. II. Title.
    III. Lecture notes in statistics (Springer-Verlag); v. 154.

HB3711 .K26 2001
388.5'42'0151955—dc21   00-059552

Printed on acid-free paper.

9 8 7 6 5 4 3 2 1

ISBN 978-0-387-95112-6   SPIN 10777073

Contents

Figures

1.
Introduction and Brief Summary

This monograph addresses the problem of measuring economic cycles (also called business cycles) in macroeconomic time series. In the decade that followed the Great Depression, economists developed an interest in the possible existence of (more or less systematic) cycles in the economy; see, for example, Haberler (1944) or Shumpeter (1939). It became apparent that in order to identify economic cycles, one had to remove from the series seasonal fluctuations, associated with short-term behavior, and the long-term secular trend, associated mostly with technological progress. Burns and Mitchell (1946) provided perhaps the first main reference point for much of the posterior research. Statistical measurement of the cycle was broadly seen as capturing the variation of the series within a range of frequencies, after the series has been seasonally adjusted and detrended. (Burns and Mitchell suggested a range of frequencies associated with cycles with a period between, roughly, two and eight years.)

Statistical methods were devised to estimate cyclical variation, and these gradually evolved towards methods fundamentally based on the application of moving average filters to the series; see, for example, Bry and Boschan (1971). These moving average filters were "ad hoc" filters, in the sense that they were fixed, independent of the particular series being analyzed; they were designed as linear "band-pass" filters, that is, as filters aimed at capturing the series variation within a certain band of frequencies. The last 20 years have witnessed methodological research on two broad fronts: the first dealt with further developments of the moving average type approach; the other was the development of more complex statistical approaches oriented towards capturing cyclical features, such as asymmetries and varying period lengths, that could not be captured with univariate linear filters. Examples of research in both directions can be found in Sims (1977), Lahiri and Moore (1991), Stock and Watson (1993), and Hamilton (1994) and (1989). Although the first approach is known to present serious limitations, the new and more sophisticated methods developed in the second approach (most notably, multivariate and nonlinear extensions) are at an early stage, and have proved still unreliable, displaying poor behavior when moving away from the sample period. Despite the fact that business cycle estimation is basic to the conduct of macroeconomic policy and to monitoring of the economy, many decades of attention have shown that formal modeling of economic cycles is a frustrating issue. As Baxter and King (1999) point out, we still face at present the same basic question "as did Burns and Mitchell fifty years ago: how should one isolate the cyclical component of an economic time series? In particular, how should one separate business-cycle elements from slowly evolving secular trends, and rapidly varying seasonal or irregular components?"

Be that as it may, it is a fact that measuring (in some way) the business cycle is an actual pressing need of economists, in particular of those related to the functioning of policy-making agencies and institutions, and of applied macroeconomic research. Lacking a practical and reliable alternative, moving average methods are the ones actually used, to the point that economic agencies (such as the OECD, the International Monetary Fund, or the European Central Bank) often have internal rules or recommendations to measure economic cycles that are moving-average-type methods. One can say that, very broadly, within the set of applied business-cycle analysts, there has been a convergence towards what could be called "Hodrick–Prescott" (HP) filtering, a methodology proposed by Hodrick and Prescott (1980); see also Kydland and Prescott (1982) and Prescott (1986). The emergence of the HP filter as a paradigm has, probably, been fostered by economic globalization and European integration, which has required a relatively high level of methodological homogeneity in order to compare countries. It is important to point out that, because seasonal variation should not contaminate the cycle, for series with a higher than annual frequency of observations, the HP filter is applied to seasonally adjusted (SA) series. Seasonal adjustment is most often performed with the program X11, also an ad hoc filter designed to remove seasonal variation (see Shiskin et al. (1967)). Therefore, the present paradigm in applied work having to do with business cycle estimation is to apply the HP filter to X11-filtered series. An unavoidable consequence is that, to some degree, the procedure eventually is often used as a black box.

Academic criticism of the HP filter has pointed out some serious drawbacks. But, beyond the criticism, not much effort has been spent on addressing these shortcomings. An important exception is Baxter and King (1999), where an alternative (related) filter is proposed that improves smoothness of the estimators for the central years, but avoids estimation at both extremes of the series (including, of course, current and recent periods). Systematic improvement of the filter performance has been clearly hampered by its ad hoc nature and the lack of an underlying statistical model with a precise definition of the components (see, e.g., Harvey (1985) and Crafts et al. (1989)). Notwithstanding the criticisms, its widespread use in practice may evidence (besides its simplicity) the empirical fact that, as a first (or rough) approximation, analysts find the results useful. In fact, in recent years HP cycles are also applied as a way of measuring the so-called output gap (i.e., the difference between potential and actual output), which is becoming central to the conduct of monetary policy and the setting of interest rates by central banks.

In this monograph, first, we analyze in detail some of the major limitations of the HP filter and of the combined X11-HP filter. By incorporating time series analysis techniques, mostly developed over the last 20 years as an aftermath of the explosion in the use of ARIMA-type models (Box and Jenkins, 1970), it is seen how some intuitive and relatively simple modifications to the filter can improve significantly its performance, in particular in terms of cleanness of the signal, smaller revision, stability of end-period estimators, and detection of turning points. The main modification we introduce is the replacement of the X11-SA series by a model-based trend-cycle estimator, extended at both ends with optimal forecasts and backcasts, as input to the HP filter, and to apply the HP filter in an ARIMA-model-based (AMB) type of algorithm, which is described in detail. A preliminary discussion of some of the topics in parts of this monograph is contained in Kaiser and Maravall (1999).

Then, we show how the modified filter can be seen as the exact solution of a well-defined statistical problem, namely, optimal (minimum mean squared error) estimation of components in a standard unobserved-component model, where the observed series is decomposed into a trend, a cycle, a seasonal, and an irregular component. This problem is straightforward to solve with the well-known Kalman or Wiener–Kolmogorov filter techniques (see, e.g., Harvey (1989) and Maravall (1995)). The models for the components incorporate some a priori features that reflect the ad hoc nature of the HP filter, and some series-dependent features that ensure compatibility with the stochastic structure of the particular series at hand, in the sense that the aggregate ARIMA model implied by the components is exactly the parsimonious ARIMA model identified directly on the observed series. Furthermore, summing the trend and the cycle, the standard trend-cycle/ seasonal/irregular AMB decomposition, originally suggested by Burman (1980) and Hillmer and Tiao (1982), is obtained. An obvious advantage of the model-based interpretation is that it greatly facilitates diagnostics and inference, thereby facilitating systematic analysis and improvement. Finally, it is shown how the procedure is trivially implemented with already available free software.

The book is structured as follows. The next chapter (Chapter 2) presents a review of some basic time series concepts and tools, and is oriented towards making the book as self-contained as possible. Chapter 3 reviews ARIMA and unobserved components models, as well as optimal estimation of the components. Attention focuses on the so-called AMB method for decomposing a time series. Chapter 4 presents and discusses the Hodrick–Prescott filter, and develops an alternative derivation of the filter, based on a Wiener–Kolmogorov-type representation, that provides an efficient computational algorithm and greatly facilitates analysis and interpretation of the filter.

Chapter 5 looks at some important limitations of the HP filter. First, the problem of imprecise endpoint estimation and large revisions in recent estimators is addressed. Second, we discuss the problem of spurious results associated with the ad hoc nature of the HP filter. Finally, the problem of noise contamination of the cyclical signal is considered. Chapter 6 presents some relatively straightforward improvements to the HP (and X11-HP) filter that can reduce significantly the previous limitations; it is seen that a side result is a clear improvement in early detection of turning points. We refer to the improved filter as the modified Hodrick–Prescott (MHP) filter.

The last chapter shows how the MHP filter can be exactly obtained from an AMB approach. In particular, it is proved that, for a large class of time series, the MHP filter is identical to the optimal (minimum mean square error) estimator of the cycle in a fully specified unobserved component model, where the other components are the seasonal, irregular, and (long-term) trend components, that, by construction, respects the stochastic structure of the series. It is further seen how the model-based framework can be exploited to answer questions of applied interest, such as what are the confidence intervals around the cycle estimator, or testing hypotheses concerning its evolution (and forecast).

The discussion of Chapters 5, 6, and 7 is illustrated in detail with an example consisting of four Spanish quarterly economic indicators.

2.
A Brief Review of Applied Time Series Analysis

2.1 Some Basic Concepts

In the introduction it was mentioned that the present standard technique used in applied work to estimate business cycles consists of applying a moving average (MA) filter, most often the HP filter, to a seasonally adjusted (SA) series, most often adjusted with the X11 filter. It was also mentioned that the procedure presents several drawbacks which, as shown, can be seriously reduced by incorporating some time series analysis tools, such as ARIMA models and signal extraction techniques. Before proceeding further, it will prove convenient to review some concepts and tools of applied time series analysis.

A previous word of caution should be said. The standard filtering procedure to estimate business cycles may require some prior corrections to the series, given that otherwise the results can be strongly distorted. An important example is outlier correction, as well as the correction for special effects that can have many different causes (legal changes, modifications in the statistical measurement procedure, etc.). This "preadjustment" of the series is briefly described in Section 3.3, where references for its methodology and its application in practice are provided, that also cover the case in which observations are missing. For the rest of the book, we assume that the series either has already been preadjusted, or that no preadjustment is needed.

Furthermore, although the discussion and the approach are also valid for other frequencies of observation, in order to simplify, we concentrate on quarterly series.

The very basic intuition behind the concept of cyclical or seasonal variation leads to the idea of decomposing a series into "unobserved components", mostly defined by the frequency of the associated variation. If x_t denotes the observed series, the simplest formulation could be

$$x_t = \sum_j x_{jt} + u_t, \tag{2.1}$$

where the variables x_{jt} denote the unobserved components, and u_t a residual effect (often referred to as the "irregular component"). In the early days, the components were often specified to follow deterministic models that could be estimated by simple regression. We follow the convention: a deterministic model denotes a model that yields forecasts with zero error when the model parameters are known. Stochastic Models will provide forecasts with non zero random errors even when the parameters are known. For example, a deterministic trend component (p_t) could be specified as the linear trend

$$p_t = a + bt, \tag{2.2}$$

and the seasonal component (s_t) could be modeled with dummy variables, as in

$$s_t = \sum_j c_j d_{jt}, \tag{2.3}$$

where $d_{jt} = 1$ when t corresponds to the jth period of the year, and $d_{jt} = 0$ otherwise. An equivalent formulation can be expressed in terms of deterministic sine-cosine functions.

Gradual realization that seasonality evolves in time (an obvious example is the weather, one of the basic causes of seasonality) leads to changes in the estimation procedure. It was found that linear filters could reproduce the moving features of a trend or a seasonal component. A linear filter simply denotes a linear combination of the series x_t, as in

$$y_t = c_{-k_1} x_{t-k_1} + \ldots + c_{-1} x_{t-1} + c_0 x_t + c_1 x_{t+1} + \ldots + c_{k_2} x_{t+k_2}, \tag{2.4}$$

and, insofar as y_t is then some sort of moving average of successive stretches of x_t, we also use the expression moving average filter. The weights c_j could be found in such a way as to capture the relevant variation associated with the particular component of interest. Thus a filter for the trend would capture the variation associated with the long-term movement of the series, and a filter for a seasonal component would capture variation of a seasonal nature. A filter designed in this way, with an "a priori" choice of the weights, is an "ad hoc" fixed filter, in the sense that it is independent of the particular series to which it is being applied. Both, the HP and the X11 filters can be seen as ad hoc fixed MA filters (although, strictly speaking, the coefficients as show later, are not constant.)

Over time, however, application of ad hoc filtering has evidenced some serious limitations. An important one is the fact that, due to its fixed character, spurious results can be obtained, and for some series the component may be overestimated, while for other series, it may be underestimated. To overcome this limitation, and in the context of seasonal adjustment, an alternative approach was suggested (around 1980) whereby the filter adapted to the particular structure of the series, as captured by its ARIMA model. The approach, known as the AMB approach, consists of two steps: an ARIMA model is obtained for the observed series, and signal extraction techniques are used to estimate the components with filters that are, in some well-defined way, optimal.

2.2   Stochastic Processes and Stationarity

The following summary is an informal review, aimed at providing some basic tools for the posterior analysis, as well as some intuition for their usefulness. Our purpose is for the presentation to be as self-contained as possible. More complete treatments of time series analysis are provided in many textbooks; some helpful references are Box and Jenkins (1970), Brockwell and Davis (1987), Granger and Newbold (1986), Harvey (1993), and Mills (1990).

The starting point is the concept of a stochastic process. For our purposes, a stochastic process is a real-valued random variable z_t, that follows a distribution $f_t(z_t)$, where t denotes an integer that indexes the period. The T-dimensional variable $(z_{t_1}, z_{t_2}, \ldots, z_{t_T})$ has a joint distribution that depends on $(t_1, t_2, \ldots, t_T)$. A time series $[z_{t_1}, z_{t_2}, \ldots, z_{t_T}]$ denotes a particular realization of the stochastic process. Thus, for each distribution f_t, there is only one observation available. Not much can be learned from this, and more structure and more assumptions need to be added. To simplify notation, we consider the joint distribution of $(z_1, z_2, \ldots, z_t)$, for which a time series is available when $t \leq T$.

From an applied perspective, the two most important added assumptions are:

Assumption A: the process is stationary;

Assumption B: the joint distribution of $(z_1, z_2, \ldots, z_t)$ is a multivariate normal distribution.

Assumption A implies the following basic condition. For any value of t,

$$f(z_1, z_2, \ldots, z_t) = f(z_{1+k}, z_{2+k}, \ldots, z_{t+k}), \tag{2.5}$$

where k is a integer; that is, the joint distribution remains unchanged if all time periods are moved a constant number of periods. In particular, letting $t = 1$, for the marginal distribution it has to be that

$$f_t(z_t) = f(z_t)$$

for every t, and hence the marginal distribution remains constant. This implies

$$E z_t = \mu_z; \qquad V z_t = V_z, \tag{2.6}$$

where E and V denote the expectation and the variance operators, and μ_z and V_z are constants that do not depend on t.

In practice, thus, stationarity implies a constant mean level and bounded deviations from it. It is a very strong requirement and few actual economic series will satisfy it. Its usefulness comes from the fact that relatively simple transformations of the nonstationary series will render it stationary. For quarterly economic series, it is usually the case that constant variance can be achieved through the log/level transformation combined with proper outlier correction, and constant mean can be achieved by differencing.

The log transformation is "grosso modo" appropriate when the amplitude of the series oscillations increases with the level of the series. As for outliers, several possible types should be considered, the most popular ones being the additive outlier (i.e., a single spike), the level shift (i.e., a step variable), and the transitory change (i.e., an effect that gradually disappears). Formal testing for the log/level transformation and for outliers are available, as well as easy-to-apply automatic procedures for doing it (see, e.g., Gómez and Maravall (2000a)). In Section 3.3 we come back to this issue; we center our attention now on achieving stationarity in mean.

2.3   Differencing

Denote by B the backward operator, such that

$$B^j z_t = z_{t-j} \qquad (j = 0, 1, 2, \ldots),$$

and let x_t denote a quarterly observed series. We use the operators:

- regular difference: $\nabla = 1 - B$;

- seasonal difference: $\nabla_4 = 1 - B^4$;

- annual aggregation: $S = 1 + B + B^2 + B^3$.

Thus $\nabla x_t = x_t - x_{t-1}$, $\nabla_4 x_t = x_t - x_{t-4}$, and $S x_t = x_t + x_{t-1} + x_{t-2} + x_{t-3}$. It is immediately seen that the three operators satisfy the identity

$$\nabla_4 = \nabla S. \tag{2.7}$$

If x_t is a deterministic linear trend, as in $x_t = a + bt$, then

$$\nabla x_t = b, \tag{2.8}$$
$$\nabla^2 x_t = 0, \tag{2.9}$$

where $\nabla^2 x_t = \nabla(\nabla x_t)$. In general, it can easily be seen that ∇^d will reduce a polynomial of degree d to a constant. Obviously, $\nabla_4 x_t$ will also cancel a constant (or reduce the linear trend to a constant); but it will also cancel other deterministic periodic functions, such as, for example, one that repeats itself every four quarters. To find the set of functions that are canceled with the transformations $\nabla_4 x_t$, we have to find the solution of the homogeneous difference equation

$$\nabla_4 x_t = (1 - B^4)x_t = x_t - x_{t-4} = 0, \tag{2.10}$$

with characteristic equation $r^4 - 1 = 0$. The solution is given by

$$r = \sqrt[4]{1},$$

that is, the four roots of the unit circle displayed in Figure 2.1. The four roots are

$$r_1 = 1, \qquad r_2 = -1, \qquad r_3 = i, \qquad r_4 = -i. \tag{2.11}$$

The first two roots are real and the last two are complex conjugates, with modulus 1 and, as seen in the figure, frequency $\omega = \pi/2$ (frequencies are always expressed in radians). Complex conjugate roots generate periodic movements of the type

$$r_t = A^t \cos(\omega t + B), \tag{2.12}$$

where A denotes the amplitude, B denotes the phase (the angle at $t = 0$), and ω the frequency (the number of full circles that are completed in one unit of time). The period of function (2.12), denoted τ, is the number of units of time it takes for a full circle to be completed, and is related to the frequency ω by the expression

$$\tau = \frac{2\pi}{\omega}. \tag{2.13}$$

Figure 2.2(a) illustrates a periodic movement of the type (2.12), with $A = 1$, $B = 0$, and $\omega = \pi/2$. From (2.11), the general solution of $\nabla_4 x_t = 0$ can be expressed as (see, e.g., Goldberg (1967))

$$x_t = c_0 + c_1 \cos\left(\frac{\pi}{2}t + d_1\right) + c_2(-1)^t,$$

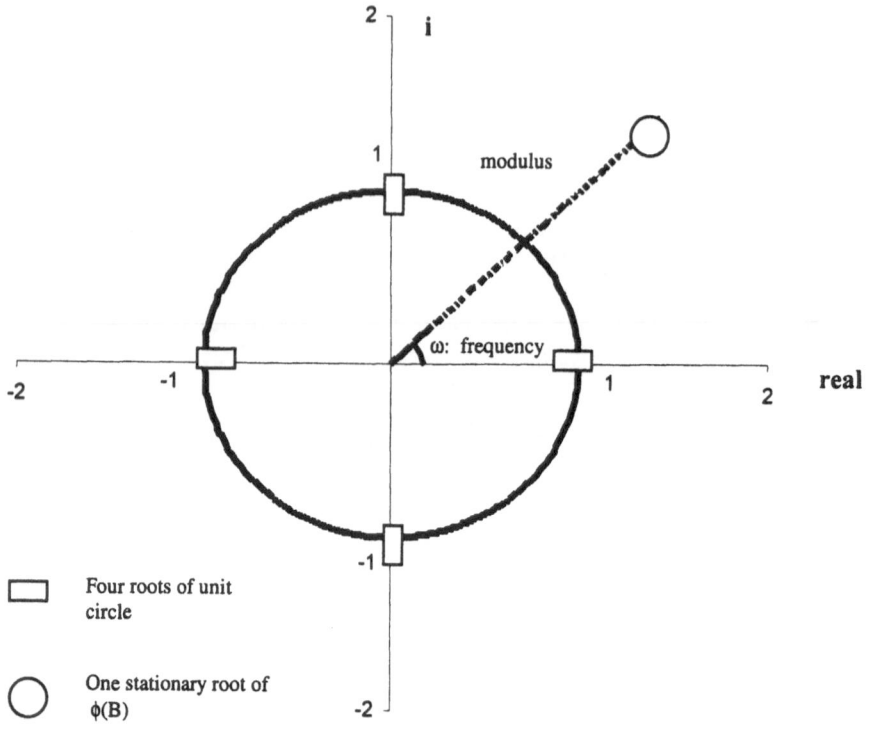

Figure 2.1. Roots of unit circle.

where c_0, c_1, c_2, and d_1 are constants to be determined from the starting conditions. Realizing that $\cos \pi = -1$, the previous expression can also be written as

$$x_t = c_0 + \sum_{j=1}^{2} c_j \cos\left(j\frac{\pi}{2}t + d_j\right), \tag{2.14}$$

with $d_2 = 0$. Considering (2.13), the first term in the sum of (2.14) is associated with a period of $\tau = 4$ quarters and thus represents a seasonal component with a once-a-year frequency; the second term has a period of $\tau = 2$ quarters, and hence represents a seasonal component with a twice-a-year frequency. The two components are displayed in Figure 2.2(b) and (c). Noticing that the characteristic equation can be rewritten as $(B^{-1})^4 - 1 = 0$, (2.11) implies the factorization

$$\nabla_4 = (1 - B)(1 + B)(1 + B^2).$$

The factor $(1 - B)$ is associated with the constant and the zero frequency, the factor $(1 + B)$ with the twice-a-year seasonality with frequency $\omega = \pi$, and the factor $(1 + B^2)$ with the once-a-year seasonality with frequency $\omega = \pi/2$. The product of these last two factors yields the annual aggregation operator S, in agreement with expression (2.7). Hence the transformation Sx_t removes seasonal nonstationarity in x_t.

For the most-often-found case in which stationarity is achieved through the differencing $\nabla\nabla_4$, the factorization

$$\nabla\nabla_4 = \nabla^2 S$$

directly shows that the solution to $\nabla\nabla_4 x_t = 0$ is of the type:

$$x_t = a + bt + \sum_{j=1}^{2} c_j \left[\cos(j\frac{\pi}{2}t) + d_j \right], \qquad (2.15)$$

with $d_2 = 0$. Thus the differencing removes the same cosine (seasonal) functions as before, plus the local linear trend (a + bt). For the case $\nabla^2\nabla_4$, the factorization $\nabla^3 S$ shows that the canceled trend is now a second order polynomial in t, the rest remaining unchanged. For quarterly series, higher-order differencing is never encountered in practice.

A final and important remark:

- Let D denote, in general, the complete differencing applied to the series x_t so as to achieve stationarity. When specifying the ARIMA model for x_t, we do not state that $Dx_t = 0$ (as, e.g., in (2.9)), but that

$$Dx_t = z_t,$$

where z_t is a zero-mean, stationary stochastic process with relatively small variance. Thus every period the solution of $Dx_t = 0$ will be perturbed by the stochastic input z_t (see Box and Jenkins (1970, Appendix A.4.1)). In terms of expression (2.15), what this perturbation implies is that the a,b,c and d coefficients will not be constant but will instead depend on time. This gradual evolution of the coefficients provides the model with an adaptive behavior that is associated with the "moving" features of the trend and seasonal components.

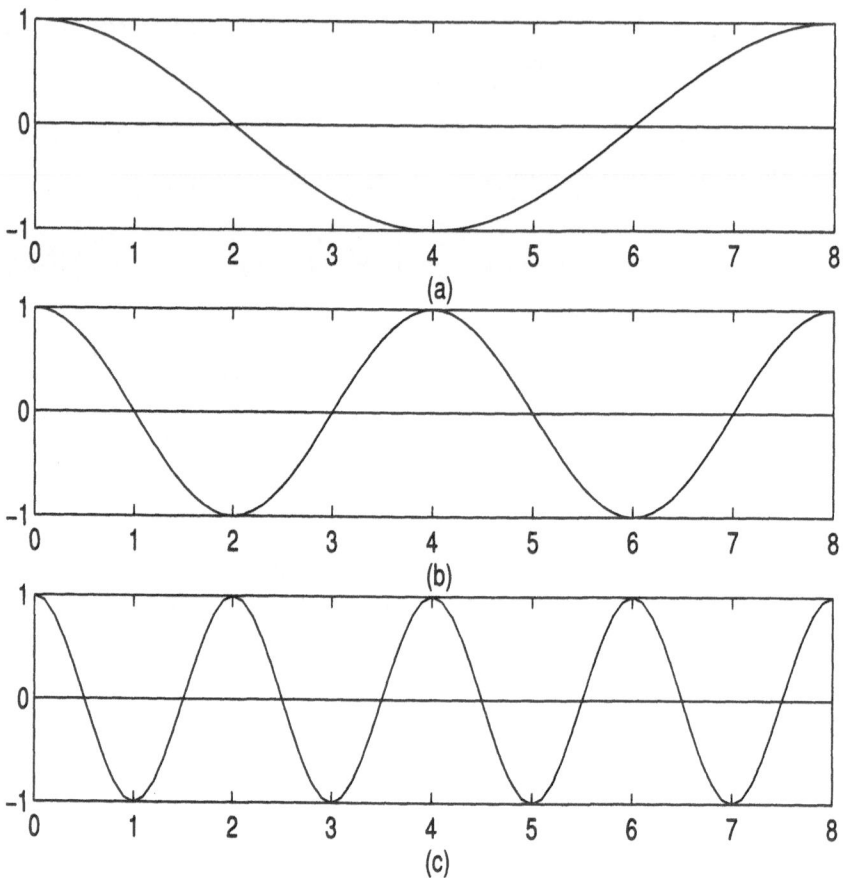

Figure 2.2. Sine-cosine functions: (a) the cosine function; (b) once-a-year frequency seasonal component; (c) twice-a-year frequency seasonal component.

2.4 Linear Stationary Process, Wold Representation, and Autocorrelation Function

Following the previous notation, if x_t denotes the observed variable and $z_t = Dx_t$ its stationary transformation, under Assumptions A and B, the variable $(z_1, z_2, \ldots, z_T)$ will have a proper multivariate normal distribution. One important property of this distribution is that the expectation of some (unobserved) variable linearly related to z_t, conditional on $(z_1, z_2, \ldots, z_T)$, will be a linear function of $z_1, z_2, \ldots, z_T$. Thus conditional expectations will directly provide linear filters. An additional important property is that, because the first two moments fully characterize the distribution, stationarity in mean and variance will imply stationarity of the process. In particular, stationarity will be implied by the constant mean and variance condition (2.6), plus the condition that

$$\mathrm{Cov}(z_t, z_{t-k}) = \gamma_k,$$

for $k = 0, \pm 1, \pm 2, \ldots$. Hence the covariance between z_t and z_{t-k} should depend on their relative distance k, not on the value of t. Therefore,

$$(z_1, z_2, \ldots, z_T) \sim N(\mu, \Sigma),$$

where μ is a vector of constant means, and Σ is the variance-covariance matrix

$$\Sigma = \begin{bmatrix} V_z & \gamma_1 & \gamma_2 & \cdots & \gamma_{T-1} \\ & V_z & \gamma_1 & \cdots & \gamma_{T-2} \\ & & \cdots & \cdots & \\ & & & V_z & \gamma_1 \\ & & & & V_z \end{bmatrix} \quad (V_z = \gamma_0),$$

a positive definite symmetric matrix. Let F denote the forward operator, $F = B^{-1}$, such that

$$F^j z_t = z_{t+j} \quad (j = 0, 1, 2, \ldots),$$

a more parsimonious representation of the second-order moments of the stationary process z_t is given by the autocovariance generating function (AGF)

$$\gamma(B, F) = \gamma_0 + \sum_{j=1}^{\infty} \gamma_j (B^j + F^j). \tag{2.16}$$

To transform this function into a scale-free function, we divide by the variance γ_0, and obtain the autocorrelation generating function (ACF),

$$\rho(B, F) = \rho_0 + \sum_{j=1}^{\infty} \rho_j (B^j + F^j), \tag{2.17}$$

where $\rho_j = \gamma_j / \gamma_0$. If the following conditions on the AGF,

1. $\rho_0 = 1$,

2. $\rho_j = \rho_{-j}$,

3. $|\rho_j| < 1$ for $j \neq 0$,

4. $\rho_j \to 0$ as $j \to \infty$, and

5. $\sum_{j=0}^{\infty} |\rho_k| < \infty$,

are satisfied, then a zero-mean, finite variance, normally distributed process is stationary. Furthermore, under the normality assumption, a complete realization of the stochastic process is fully characterized by μ_z, V_z, and $\rho(B, F)$.

When $\rho_j = 0$ for all $j \neq 0$, the process is denoted a white noise process. Therefore, a white noise process is a sequence of normally identically independently distributed random variables.

The AGF (or ACF) is the basic tool in the so-called "time domain analysis" of a time series. The first statistics that we compute for a time series $[z_1, \ldots, z_T]$ are estimates of the autocovariances and autocorrelations using the standard sample estimates

$$\bar{z} = T^{-1} \sum_{t=1}^{T} z_t; \qquad \hat{\gamma}_k = T^{-1} \sum_{t=k+1}^{T} (z_t - \bar{z})(z_{t-k} - \bar{z}); \qquad \hat{\rho}_k = \hat{\gamma}_k / \hat{\gamma}_0.$$

Next, a look at the sample ACF (SACF) gives an idea of the lag dependence in the series: large autocorrelation for low lags points towards large inertia; large autocorrelation for seasonal lags, of course, indicates the presence of seasonality. One word of caution should be nevertheless made: the dependence of the autocorrelation estimators on the same time series can induce important spurious correlation between them. These correlations can have serious distorting effects on the visual aspect of the SACF, which may fail to damp out according to expectations (see Box and Jenkins (1970, section 2.1)). Figure 2.3(a) exhibits the ACF of a quarterly stationary process; Figure 2.3(b) displays the SACF obtained with a sample of 100 observations. As a consequence, care should be taken not to "over-read" SACFs, ignoring large-lag autocorrelations, and focusing only on their most salient features.

To start the modeling procedure, a general result on linear time series processes provides us with an analytical representation of the process that proves very useful. This is the so-called Wold (or fundamental) representation. We present it next.

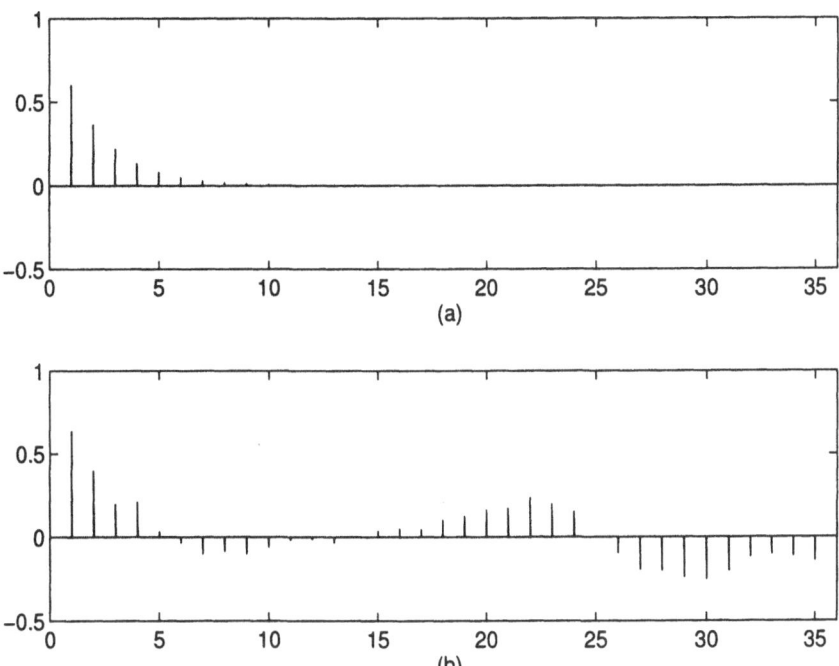

Figure 2.3. (a) Theoretical ACF; (b) sample ACF.

Let z_t denote a linear stationary stochastic process with no deterministic component; then z_t can be expressed as the one-sided moving average

$$
\begin{aligned}
z_t &= a_t + \psi_1 a_{t-1} + \psi_2 a_{t-2} + \dots \\
&= \sum_{j=0}^{\infty} \psi_j a_{t-j} = \Psi(B) a_t, \\
\Psi(B) &= \sum_{j=0}^{\infty} \psi_j B^j \qquad (\psi_0 = 1),
\end{aligned}
\qquad (2.18)
$$

where a_t is a white noise process with zero mean and constant variance V_a, and $\Psi(B)$ is such that

1. $\psi_j \to 0$ as $j \to \infty$,

2. $\sum_{j=0}^{\infty} |\psi_j| < \infty$,

the last condition reflecting a sufficient condition for convergence of the polynomial $\Psi(B)$. Given the ψ_j-coefficients, a_t represents the one-period-ahead forecast error of z_t; that is,

$$a_t = z_t - \hat{z}_{t|t-1},$$

where $\hat{z}_{t|t-1}$ is the forecast of z_t made at period t - 1. Since a_t represents what is new in z_t, that is, what is not contained in its past $[z_{t-1}, z_{t-2}, z_{t-3}, \ldots]$, it is referred to as the innovation of the process. The representation of z_t in terms of its innovations, given by (2.18), is unique, and is usually referred to as the Wold representation.

A useful result is the following: if $\gamma(B, F)$ represents the AGF of the process z_t, then

$$\gamma(B, F) = \Psi(B)\Psi(F)V_a. \tag{2.19}$$

In particular, for the variance,

$$V_z = (1 + \psi_1^2 + \psi_2^2 + \ldots)V_a. \tag{2.20}$$

2.5   The Spectrum

The spectrum is the basic tool in the so-called "frequency domain approach" to time series analysis. It represents an alternative way to view and interpret the information contained in the second-order moments of the series. The frequency approach is particularly convenient for analyzing unobserved components, such as trends, cycles, or seasonality. Our aim is not to present a complete and rigorous description, but to provide some intuition and basic understanding that will permit us to use it properly for our purposes. (Two good references for a general presentation are Jenkins and Watts (1968) and Grenander and Rosenblatt (1957).)

Consider, first, a time series (i.e., a partial realization of a stochastic process) given by $z_1, z_2, \ldots, z_T$. To simplify the discussion, assume the process has zero mean and that T is even, so that we can write $T = 2q$. In the same way that, as is well known, the T values of z_t can be exactly duplicated ("explained") by a polynomial of order $(T$ - $1)$, they can also be exactly reproduced as the sum of $T/2$ cosine functions of the type (2.12); this result provides in fact the basis of Fourier analysis.

Figure 2.4(a) shows, for example, the quarterly time series of 10 observations generated by the five cosine functions of Figure 2.4(b). To construct this set of functions, we start by defining the fundamental frequency $\omega = 2\pi/T$ (i.e., the frequency of one full circle completed in T periods) and its multiples (or harmonics) $\omega_j = (2\pi/T)j$, $j = 1, 2, \ldots, q$. Then, express (2.12) as

$$r_{jt} = a_j \cos \omega_j t + b_j \sin \omega_j t, \tag{2.21}$$

and hence,

$$z_t = \sum_{j=1}^{q} r_{jt}. \tag{2.22}$$

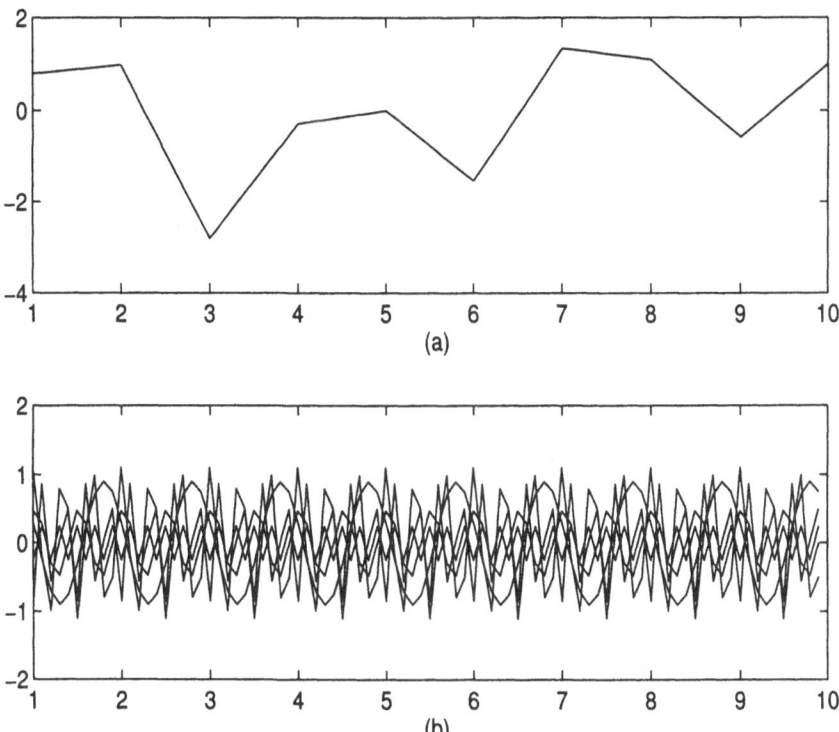

Figure 2.4. (a) Generated time series; (b) Fourier series.

It is straightforward to check that a_j and b_j are related to the amplitude A_j by $A_j^2 = a_j^2 + b_j^2$. From (2.21) and (2.22), by plugging in the values of z_t, w_j, and t, a linear system of T equations is obtained in the unknown a_js and b_js, $j = 1, 2, \ldots, q$; a total of T unknowns. Therefore, for each frequency w_j, we obtain a square amplitude A_j^2. The plot of A_j^2 versus w_j, $j = 1, \ldots, q$, is the periodogram of the series.

As a consequence, we obtain a set of periodic functions with different frequencies and amplitudes. We can group the functions in intervals of frequency by summing the squared amplitudes of the functions that fall in the same interval. In this way we obtain a histogram of frequencies that shows the contribution of each interval of frequency to the series variation; an example is shown in Figure 2.5(a). In the same way that a density function is the model counterpart of the usual histogram, the spectrum is the model counterpart of the frequency histogram (properly standardized).

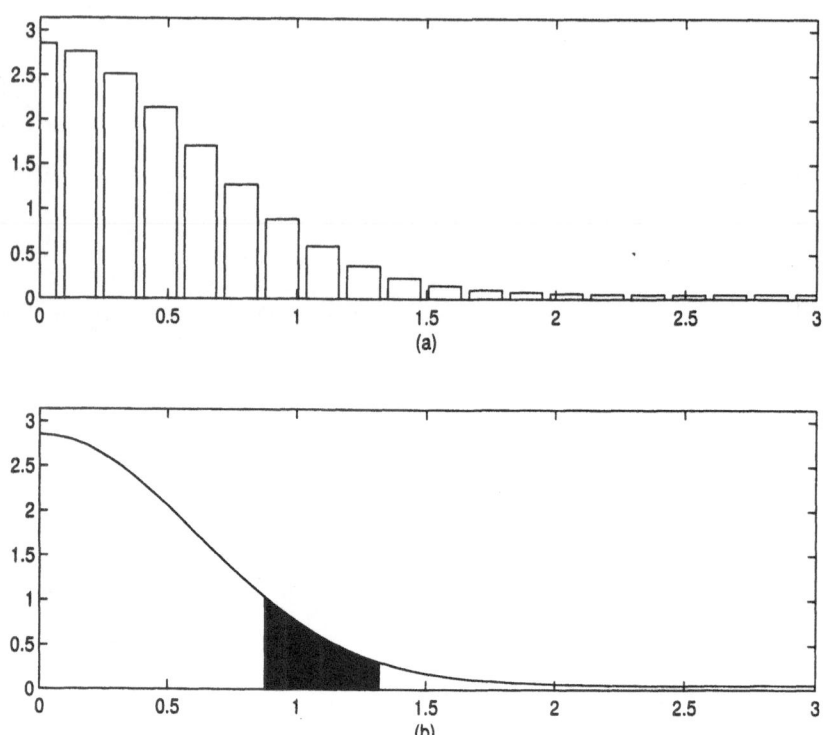

Figure 2.5. (a) Histogram of frequencies; (b) power spectrum.

We can now let the interval $\Delta\omega_j$ go to zero, and the frequency histogram becomes a continuous function, which is denoted the sample spectrum. The area over the differential $d\omega$ represents the contribution of the frequencies in $d\omega$ to the variation of the time series. An important result links the sample spectrum with the SACF (see Box and Jenkins (1970, Appendix A.2.1)). If $H(\omega)$ denotes the sample spectrum, then it is proportional to

$$H(\omega) \propto \left(\hat{\gamma}_0 + 2 \sum_{t=1}^{T-1} \hat{\gamma}_j \cos \omega t \right), \tag{2.23}$$

where $\hat{\gamma}_j$ denotes the lag-j autocovariance estimator.

The model equivalent of (2.23) provides precisely the definition of a power spectrum. Consider the AGF of the stationary process z_t, given by

$$\gamma(B, F) = \gamma_0 + \sum_{j=1}^{\infty} \gamma_j (B^j + F^j), \tag{2.24}$$

where B is a complex number of unit modulus, which can be expressed as $e^{i\omega}$. Replacing B and F by their complex representation, (2.24) becomes the function

$$g(\omega) = \gamma_0 + \sum_{j=1}^{\infty} \gamma_j (e^{-i\omega j} + e^{i\omega j}),$$

or, using the identity $[e^{-i\omega j} + e^{i\omega j} = 2\cos(j\omega)]$, and dividing by 2π, one obtains

$$g_1(\omega) = \frac{1}{2\pi} \left[\gamma_0 + 2 \sum_{j=1}^{\infty} \gamma_j \cos(j\omega) \right]. \tag{2.25}$$

The move from (2.24) to (2.25) is the so-called Fourier cosine transform of the AGF $\gamma(B, F)$, and is denoted the power spectrum. Replacing the AGF by the ACF (i.e., dividing by the variance γ_0), we obtain the spectral density function

$$g_1^*(\omega) = \frac{1}{2\pi} \left[1 + 2 \sum_{j=1}^{\infty} \rho_j \cos(j\omega) \right]. \tag{2.26}$$

It is easily seen that $g_1(\omega)$ —or $g_1^*(\omega)$— are periodic functions, and hence the range of frequencies can be restricted to $(-\pi, \pi)$ or $(0, 2\pi)$. Moreover, given that the cosine function is symmetric around zero, we only need to consider the range $(0, \pi)$. It is worth mentioning that the sample spectrum (2.23), divided by 2π, is also the Fourier transform of the sample autocovariance function.

From (2.25), knowing the AGF of a process, the power spectrum is trivially obtained. Alternatively, knowledge of the power spectrum permits us to derive the AGF by means of the inverse Fourier transform, given by

$$\gamma_k = \int_{-\pi}^{\pi} g(\omega) \cos(\omega k) d\omega.$$

Thus, for $k = 0$,

$$\gamma_0 = \int_{-\pi}^{\pi} g(\omega) d\omega, \tag{2.27}$$

which shows that the integral of the power spectrum is the variance of the process. Therefore, the area under the spectrum for the interval $d\omega$ is the contribution to the variance of the series that corresponds to the range of frequencies $d\omega$ (as in Figure 2.5(b)). Roughly, the power spectrum can be seen as a decomposition of the variance by frequency.

For the rest of the monograph, in order to simplify the notation, power spectra are expressed in units of 2π, and, because of the symmetry condition, only the range $\omega \in [0, \pi]$ is considered. We refer to this function simply as the spectrum.

As an example, consider a process z_t, the output of the second-order homogeneous difference-equation (deterministic) model

$$z_t + .81 z_{t-2} = 0 \qquad (2.28)$$

The characteristic equation $r^2 + .81 = 0$ yields the pair of complex conjugate numbers $r = \pm .9i$; situated in the imaginary axis, they are associated thus with the frequency $\omega = \pi/2$ (see Figure 2.1). The process follows therefore the deterministic function

$$z_t = .9 \cos\left(\frac{\pi}{2} t + \beta\right), \qquad (2.29)$$

where we can set $\beta = -\pi/2$. The function (2.29) does not depend on ω and the movements of z_t are all associated with the single frequency $\omega = \pi/2$. This explains the isolated spike for that frequency in Figure 2.6(a). To transform the previous model into a stochastic process, every period the equilibrium (2.28) is perturbed with a white noise (0,1) variable a_t, so that it is replaced by the stochastic model

$$z_t + .81 z_{t-2} = a_t \quad \text{or} \quad (1 + .81 B^2) z_t = a_t. \qquad (2.30)$$

From (2.30), the Wold representation (2.18) is immediately obtained as

$$z_t = \frac{a_t}{1 + .81 B^2},$$

with $\Psi(B) = 1/(1 + .81 B^2)$. Using (2.19), the AGF of z_t can be obtained through

$$
\begin{aligned}
\gamma(B, F) &= \frac{V_a}{(1 + .81 B^2)(1 + .81 F^2)} \\
&= \frac{V_a}{1.656 + .81(B^2 + F^2)}.
\end{aligned}
$$

Replacing $(B^2 + F^2)$ by $2\cos 2\omega$, the spectrum is found to be equal to

$$g(\omega) = \frac{V_a}{1.656 + 1.62 \cos 2\omega}; \qquad 0 \leq \omega \leq \pi.$$

The spike of the previous case, as seen in Figure 2.6(b)b, has now become a hill. If we increase the variance of the stochastic input a_t, as shown in part (c)c of the figure, the width of the hill (i.e., the dispersion of ω around $\pi/2$) increases. Figure 2.7 compares the type of movements generated in the three cases. As the variance of the stochastic input becomes larger, the component becomes less stable and more "moving".

In summary, if a series contains an important component for a certain frequency ω_0, its spectrum should reveal a peak around that frequency.

Given that a good definition of a trend is a cyclical component with period $\tau = \infty$, the spectral peak in this case should occur at the frequency $\omega = 0$.

To see some examples of spectra for some simple processes, we use the previous result that allows us to move from the Wold representation to the AGF, and from the AGF to the spectrum. The sequence is, in all cases,

$$
\begin{aligned}
z_t &= \Psi(B)a_t : &&\text{Wold representation ;}\\
\gamma(B, F) &= \Psi(B)\Psi(F)V_a : &&\text{AGF of } z_t\\
&= [\gamma_0 + \sum_j \gamma_j(B^j + F^j)]V_a;\\
g(\omega) &= [\gamma_0 + 2\sum_j \gamma_j \cos j\omega]V_a : &&\text{spectrum.}
\end{aligned}
$$

1. *White noise process.* Then, $\gamma_j = 0$ for $j \neq 0$, and hence

$g(\omega) = $ constant (Figure 2.8(a)).

2. *Moving average process of order 1: MA(1)*

$z_t = a_t + \theta_1 a_{t-1}$

$z_t = (1 + \theta_1 B)a_t$, hence $\Psi(B) = (1 + \theta_1 B)$; therefore

$$
\begin{aligned}
\gamma(B, F) &= \Psi(B)\Psi(F)V_a = (1 + \theta B)(1 + \theta F)V_a =\\
&= [1 + \theta^2 + \theta(B + F)]V_a,
\end{aligned}
$$

$g(\omega) = [1 + \theta^2 + 2\theta \cos \omega]V_a.$

Figure 2.8(b) shows an example with $\theta < 0$.

3. *Autoregressive process of order 1: AR(1)*

$z_t + \phi_1 z_{t-1} = a_t;$ or $(1 + \phi B)z_t = a_t$

$z_t = (1/(1 + \phi B)) a_t$, so that $\Psi(B) = 1/(1 + \phi B)$;

assuming $|\phi| < 1$, it is found that

$$
\begin{aligned}
\gamma(B, F) &= [(1 + \phi B)(1 + \phi F)]^{-1} V_a\\
&= [1 + \phi^2 + \phi(B + F)]^{-1} V_a;
\end{aligned}
$$

$g(\omega) = [1 + \phi^2 + 2\phi \cos \omega]^{-1} V_a.$

The case $\phi < 0$ is displayed in Figure 2.8(c). The spectrum consists of a peak for $\omega = 0$ that decreases monotonically in the range $[0, \pi]$. Therefore, the AR(1) process in this case reveals a trend-type behavior.

Figure 2.8(c) also displays (dotted line) the case $\phi > 0$. The resulting spectrum is symmetric to the previous one around the frequency $\omega = \pi/2$, and, consequently, displays a peak for $\omega = \pi$. The period associated with that peak is, according to (2.13), always 2. Therefore the AR(1) in this case reveals a cyclical behavior with period $\tau = 2$. If the data are monthly, this behavior corresponds to the six-times-a-year seasonal frequency; for a quarterly time series, to the twice-a-year seasonal frequency; for annual data, it would represent a two-year cycle effect.

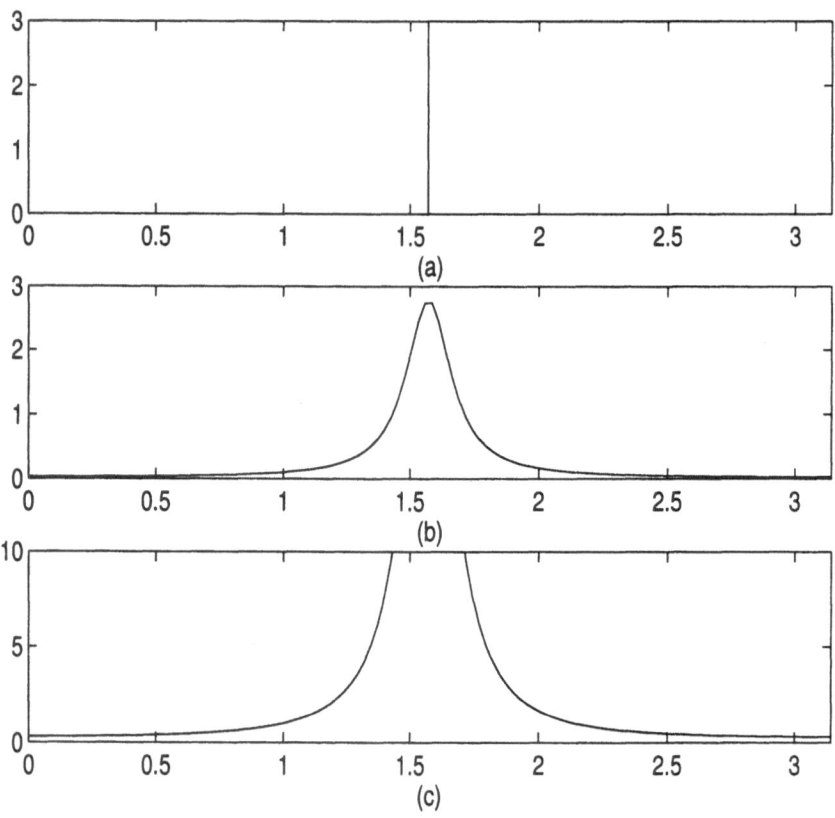

Figure 2.6. Spectra of AR(2) process: (a) deterministic component; (b) stable stochastic component; (c) highly stochastic component.

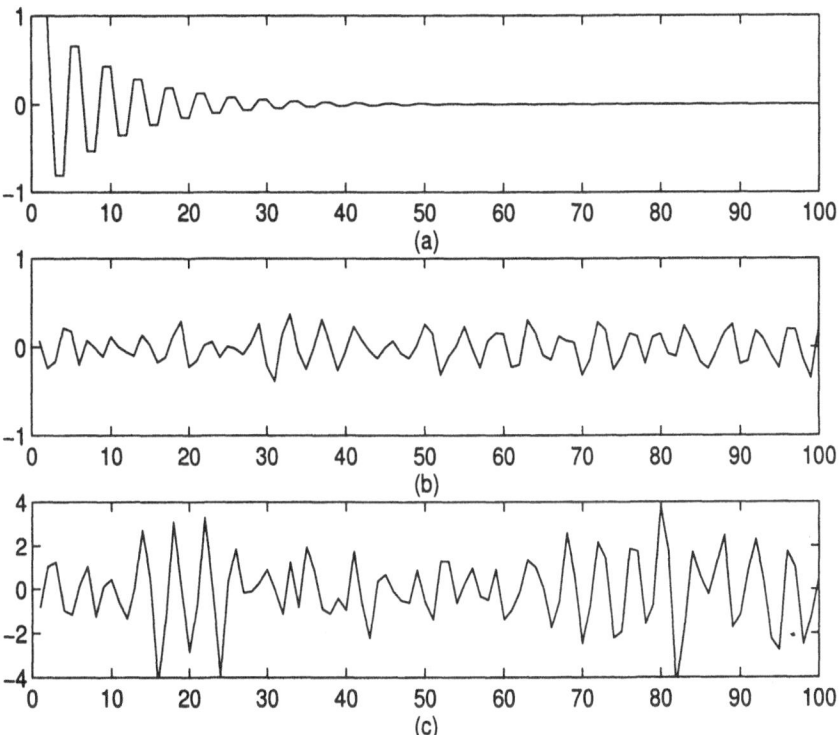

Figure 2.7. Realization of AR(2) Process: (a) deterministic component; (b) stable stochastic component; (c) highly stochastic component.

4. *Autoregressive process of order 2*: AR(2)

$$z_t + \phi_1 z_{t-1} + \phi_2 z_{t-2} = a_t \tag{2.31}$$

or

$$(1 + \phi_1 B + \phi_2 B^2) z_t = a_t. \tag{2.32}$$

Concentrating, as we did earlier, on the homogeneous part of (2.31), the characteristic equation associated with that part is precisely the polynomial in B, with $B = r^{-1}$. Thus we can find the dominant behavior of z_t from the solution of $r^2 + \phi_1 r + \phi_2 = 0$. Two cases can happen:

(a) the two roots are real;

(b) the two roots are complex conjugates.

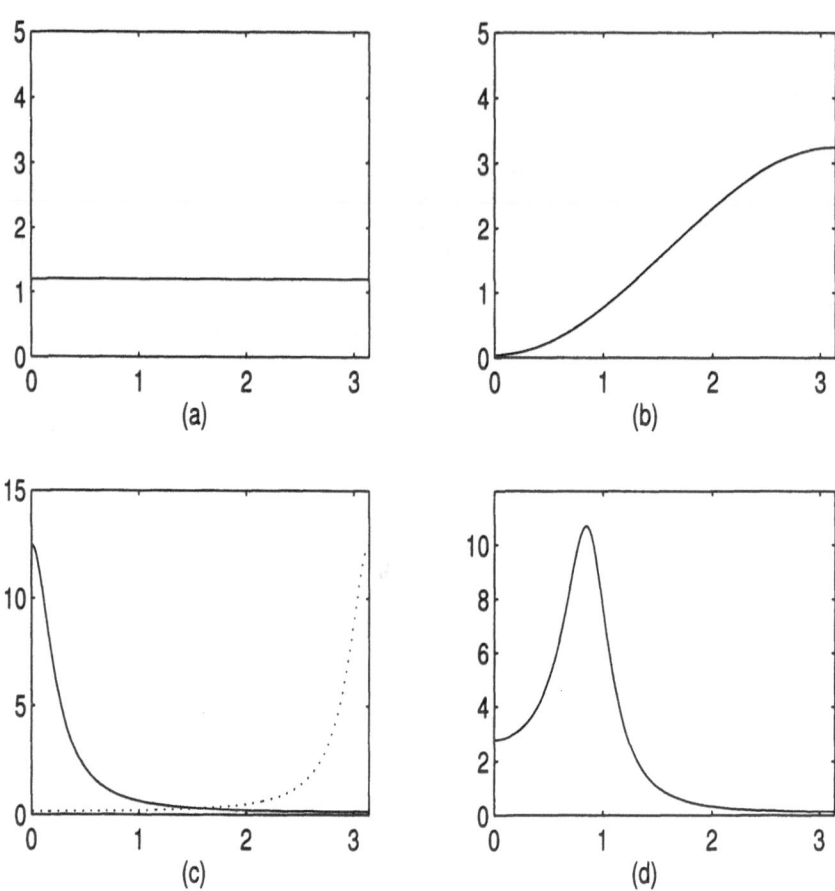

Figure 2.18. Examples of spectra: (a) spectrum of white noise; (b) spectrum of MA(1); (c) Spectrum of AR(1); (d) spectrum of AR(2).

In Case (a), if r_1 and r_2 are the two roots (we assume $|r_1|$ and $|r_2|$ are < 1), the polynomial can be factorized as $(1 - r_1B)(1 - r_2B)$, and each factor will produce the effect of an AR(1) process. Thus, if both r_1 and r_2 are > 0, the spectrum will display a peak for $\omega = 0$; if one is > 0 and the other < 0, the spectrum will have peaks for $\omega = 0$ and $\omega = \pi$; if both roots are < 0, the spectrum will have a peak for $\omega = \pi$.

In Case (b), the complex conjugate roots will generate a cosine-type (cyclical) behavior. The modulus m and the frequency ω can be obtained from the model (2.31) through

$$m = \sqrt{\phi_2}; \qquad\qquad \omega = \arccos\left(\frac{\phi_1}{2m}\right); \qquad (2.33)$$

and the spectrum will display a peak for the frequency ω, as in Figure 2.8(d).

In general, a useful way to look at the structure of an autoregressive process of order p, AR(p), a specification very popular in econometrics, is to factorize the full AR polynomial. Real roots will imply spectral peaks of the type 2.8(c), while complex conjugate roots will produce peaks of the type 2.8(d).

The range of cyclical frequencies

As already mentioned, the periodic and symmetric character of the spectrum permits us to consider only the range of frequencies $[0, \pi]$. When $\omega = 0$, the period $\tau \to \infty$, and the frequency is associated with a trend. When $\omega = \pi/2$, the period equals four quarters and the frequency is associated with the first seasonal harmonic (the once-a-year frequency). For a frequency in the range $[0 + \epsilon_1, \pi/2 - \epsilon_2]$, with $\epsilon_1, \epsilon_2 > 0$ and $\epsilon_1 < \pi/2 - \epsilon_2$, the associated period will be longer than a year, and bounded. Economic cycles should thus have a spectrum concentrated in this range. We refer to this range broadly as the "range of cyclical frequencies".

Frequencies in the range $[\pi/2, \pi]$ are associated with periods between four and two quarters. Therefore, they imply very short-term movements (with the cycle completed in less than a year) and are of no interest for business-cycle analysis. Given that $\omega = \pi$ is a seasonal frequency (the twice-a-year seasonal harmonic), the open interval of frequencies $(\pi/2, \pi)$, excluding the two seasonal frequencies, is referred to as the "range of intraseasonal frequencies".

The determination of ϵ_1 and ϵ_2 in order to specify the precise range of cyclical frequencies is fundamentally subjective, and depends on the purpose of the analysis. For quarterly data and business-cycle analysis in the context of short-term economic policy, obviously a cycle of period 100,000 years should be included in the trend, not in the business cycle. The same consideration would apply to a 10,000-year cycle. As the period decreases (and ϵ_1 becomes bigger), we eventually approach frequencies that can be of interest for business-cycle analysis. For example, if the longest cycle that should be considered is a 10-year cycle (40 quarters), from (2.13), ϵ_1 should be set as $.05\pi$.

At the other extreme of the range, very small values of ϵ_2 can produce cycles with, for example, a period of 1.2 years, too short to be of cyclical interest. If the minimum period for a cycle is set as 1.5 years, then ϵ_2 should be set equal to $.167\pi$, and the range of cyclical frequencies would be $[.05\pi, .33\pi]$. Figure 2.9 shows how, from the decision on what is the relevant interval for the periods in a cyclical component, the range of cyclical frequencies is easily determined (in the figure, the interval for the period goes from 2 to 12 years).

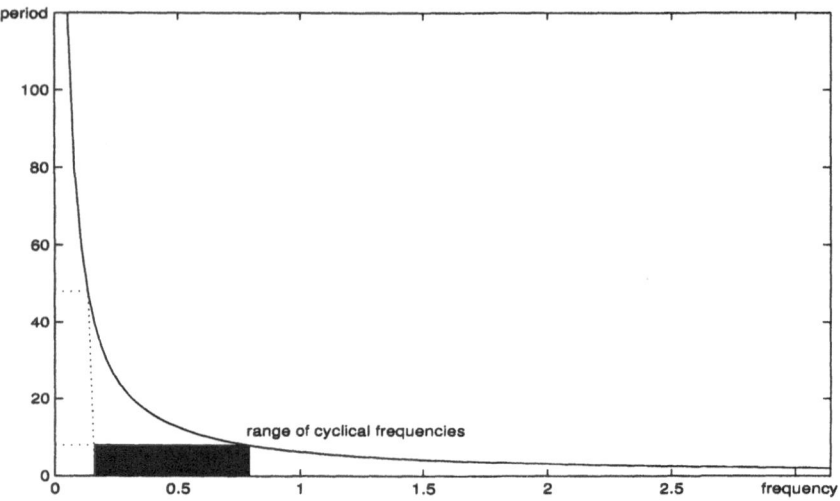

Figure 2.9. Cyclical period and frequency.

Extension to nonstationary unit roots

In the AR(1) model, we can let ϕ approach the value $\phi = -1$. In the limit we obtain

$$(1 - B)z_t = a_t \qquad \text{or} \qquad \nabla z_t = a_t,$$

the popular random walk model. Proceeding as in Case 3 in page 21, one obtains

$$g(\omega) = \frac{1}{2(1 - \cos\omega)} V_a.$$

For $\omega = 0$, $g(\omega) \to \infty$, and hence the integral (2.27) does not converge, which is in agreement with the well-known result that the variance of a random walk is unbounded. The nonstationarity induced by the root $\phi = -1$ in the AR polynomial $(1 + \phi B)$, a unit root associated with the zero frequency, induces a point of infinity in the spectrum of the process for that frequency. This result is general: a unit AR root, associated with a particular frequency ω_0, will produce an ∞ in the spectrum for that particular frequency.

An important example is when the polynomial $S = 1 + B + B^2 + B^3$ is present in the AR polynomial of the series. Given that S factorizes into $(1 + B)(1 + B^2)$, its roots are -1, and $\pm i$, associated with the frequencies π and $\pi/2$, respectively (as seen in Section 2.3). The Fourier transform of S, given by

$$S^* = 4(1 + \cos \omega)(1 + \cos 2\omega),$$

displays zeros for $\omega = \pi$ (first factor), and $\omega = \pi/2$ (second factor). Because S^* will appear in the denominator of the spectrum, its zeros will induce points of ∞. Therefore, a model with an AR polynomial including S will have a spectrum with points of ∞ for the frequencies $\omega = \pi/2$, and $\omega = \pi$, that is, the seasonal frequencies.

It follows that, in the usual case of a seasonal quarterly series, for which a $\nabla \nabla_4$ or a $\nabla^2 \nabla_4$ differencing has been used as the stationary transformation, the spectrum of the series would present points of ∞ for the frequencies $\omega = 0$, $\omega = \pi/2$, and $\omega = \pi$. Figure 2.10(a) exhibits what could be the spectrum of a standard, relatively simple quarterly series.

One final point: given that a spectrum with points of ∞ has a nonconvergent integral, and that no standardization can provide a proper spectral density, the term spectrum is usually replaced by pseudospectrum (see, e.g., Hatanaka and Suzuki (1967), and Harvey (1989)). For our purposes, however, the points of ∞ pose no serious problem, and the pseudospectrum can be used in much the same way as the stationary spectrum (this becomes clear throughout the discussion). In particular, if, for the nonstationary series, we use the nonconvergent representation (2.18), compute the function $\gamma(B, F)$ through (2.19), and, in the line of Hatanaka and Suzuki, refer to this function as the "pseudo-AGF", the pseudospectrum is the Fourier transform of the pseudo-AGF. Bearing in mind that, when referring to nonstationary series, the term "pseudospectrum" would be more appropriate, in order to avoid excess notation, we simply use the term spectrum in all cases.

2.6   Linear Filters and Their Squared Gain

Back to the linear filter (2.4) of Section 2.1, the filter can be rewritten as

$$y_t = C(B, F)x_t, \tag{2.34}$$

where

$$C(B, F) = \sum_{j=1}^{k_1} c_{-j} B^j + c_0 + \sum_{j=1}^{k_2} c_j F^j.$$

If $k_1 = k_2$ and $c_j = c_{-j}$ for all j values, the filter becomes centered and symmetric, and we can express it as

$$C(B, F) = c_0 + \sum_{j=1}^{k} c_j (B^j + F^j). \tag{2.35}$$

Using the same Fourier transform as with expression (2.24), that is, replacing $(B^j + F^j)$ by $(2 \cos j\omega)$, the frequency domain representation of the filter becomes

$$C^*(\omega) = c_0 + 2 \sum_{j=1}^{k} c_j \cos(j\omega). \tag{2.36}$$

If $k_1 \neq k_2$ or $c_j \neq c_{-j}$, the uncentered or asymmetric filter does not accept an expression of the type (2.36). Additional terms involving imaginary numbers that do not cancel out will be present. This feature will induce a phase effect in the output, in the sense that there will be a systematic distortion in the timing of events between input and output (e.g., in the dating of turning points, of peaks and throughs, etc.). For our purposes, this is a disturbing feature and hence we concentrate attention on centered and symmetric filters.

Being $C(B,F)$ symmetric and x_t stationary, (2.34) directly yields

$$AGF(y) = [C(B, F)]^2 ACF(x),$$

so that, applying the Fourier transform, we obtain

$$g_y(\omega) = [G(\omega)]^2 g_x(\omega), \tag{2.37}$$

where $g_x(\omega)$ and $g_y(\omega)$ are the spectra of the input and output series x_t and y_t and we represent by $G(\omega)$ the Fourier transform of $C(B, F)$. The function $G(\omega)$ is denoted the gain of the filter. From the relationship (2.37), the squared gain determines the contribution of the variance of the input in explaining the variance of the output for each different frequency. If $G(\omega) = 1$, the full variation of x for that frequency is passed to y; if $G(\omega) = 0$, the variation of x for that frequency is fully ignored in the computation of y.

When interest centers in the components of a series, where the components are fundamentally characterized by their frequency properties, the squared gain function becomes a fundamental tool, since it tells us which frequencies will contribute to the component and which frequencies will not enter it. As an example, consider a quarterly series with spectrum that of Figure 2.10(a). The peaks for $\omega = 0, \pi/2,$ and π imply that the series contains a trend component and a seasonal component, associated with the once-and twice-a-year frequencies. A seasonal adjustment filter is one with a squared gain displaying holes for the seasonal frequencies that removes the seasonal spectral peaks, leaving the rest basically unchanged (Figure 2.10(b) displays the squared gain of the default X11 seasonal adjustment filter). A detrending filter is one with a squared gain that removes the spectral peak for the zero frequency, and leaves the rest approximately unchanged (Figure 2.10(c) displays the squared gain of the Hodrick–Prescott detrending filter, for the case of $\lambda = 1000$).

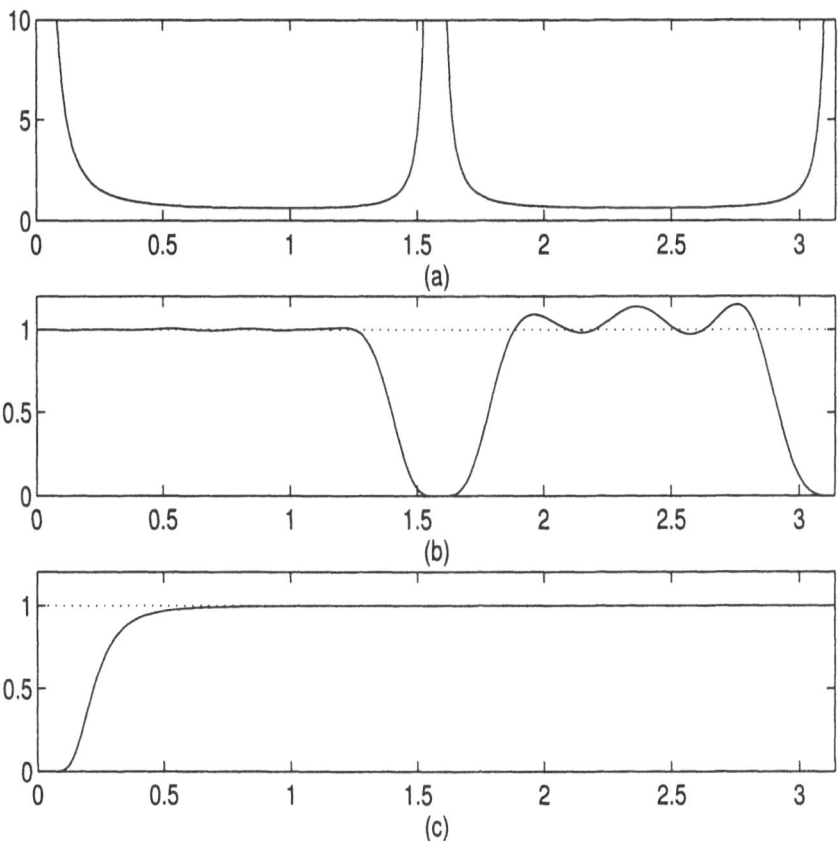

Figure 2.10. (a) Series spectrum; (b) gain of seasonal filter (default X11); (c) gain of a detrending filter (HP with $\lambda = 1000$).

One final important clarification should be made. We said that, in order to avoid phase effects, symmetric and centered filters are considered. Let one such filter be

$$y_t = c_k x_{t-k} + \ldots + c_1 x_{t-1} + c_0 x_t + c_1 x_{t+1} + \ldots + c_k x_{t+k}. \qquad (2.38)$$

Assume a long series and let T denote the last observed period. When $T \geq t + k$, the filter can be applied to obtain y_t with no problem. However, when $T < t + k$, observations at the end of the series, needed to compute y_t, are not available yet, and hence the filter cannot be applied. As a consequence, the series y_t cannot be obtained for recent enough periods, because unknown future observations of x_t are needed. The fact that interest typically centers on recent periods has led filter designers to modify the weights of the filters when truncation is needed because of a lack of future observations (see, e.g., the analysis in Burridge and Wallis (1984), in the context of the seasonal adjustment filter X11). Application of these truncated filters yields a preliminary measure of y_t, because new observations will imply changes in the weights, until $T \geq t + k$ and the final (or historical) value of y_t can be obtained. One modification that has become popular is to replace needed future values, not yet observed, by their optimal forecasts, often computed with an ARIMA model for the series x_t. Given that the forecasts are linear functions of present and past values of x_t, the preliminary value of y_t obtained with the forecasts will be a truncated filter applied to the observed series. Naturally, preliminary (truncated) filters will not be centered, nor symmetric. (In particular, the measurement of y_t obtained when the last observed period is t, i.e., when $T = t$, the so-called "concurrent" estimator, will be a purely one-sided filter.) Besides its natural appeal, replacing unknown future values with optimal forecasts has the convenient features of minimizing (within the limitations of the structure of the particular series at hand) both the phase effect and the size of the total revision the preliminary estimator will undergo until it becomes final. To this important issue of preliminary estimation and revisions we return in the following chapters.

3.

ARIMA Models and Signal Extraction

3.1 ARIMA Models

Back to the Wold representation (2.18) of a stationary process, $z_t = \Psi(B)a_t$, the representation is of no help from the point of view of fitting a model because, in general, the polynomial $\Psi(B)$ will contain an infinite number of parameters. Therefore we use a rational approximation of the type

$$\Psi(B) \doteq \frac{\theta(B)}{\phi(B)},$$

where $\theta(B)$ and $\phi(B)$ are finite polynomials in B of order q and p, respectively. Then we can write

$$z_t = \frac{\theta(B)}{\phi(B)}a_t, \text{ or}$$

$$\phi(B)z_t = \theta(B)a_t. \tag{3.1}$$

The model

$$(1 + \phi_1 B + \ldots + \phi_p B^p)z_t = (1 + \theta_1 B + \ldots + \theta_q B^q)a_t \tag{3.2}$$

is the autoregressive moving average process of orders p and q; in brief, the ARMA(p,q) model. For further reference, the inverse model of (3.1) is the one that results from interchanging the AR and MA polynomials. Thus

$$\theta(B)y_t = \phi(B)b_t,$$

with b_t white noise, is an inverse model of (3.1). Equation (3.2) can be seen as a nonhomogeneous difference equation with forcing function $\theta(B)a_t$, an MA(q) process. Therefore, if both sides of (3.2) are multiplied by z_{t-k}, with $k > q$, and expectations are taken, the right-hand side of the equation vanishes, and the left-hand side becomes:

$$\gamma_k + \phi_1\gamma_{k-1} + \ldots + \phi_p\gamma_{k-p} = 0, \tag{3.3}$$

or

$$\phi(B)\gamma_k = 0, \tag{3.4}$$

where B operates on the subindex k. The eventual autocorrelation function (i.e., γ_k as a function of k, for $k > q$) is the solution of the homogeneous difference equation (3.3), with characteristic equation

$$r^p + \phi_1 r^{p-1} + \ldots + \phi_p = 0. \tag{3.5}$$

If $r_1, \ldots, r_p$ are the roots of (3.5) the solution of (3.3) can be written as

$$\gamma_k = \sum_{i=1}^{p} r_i^k,$$

and will converge to zero as $k \to \infty$ when $|r_i| < 1, i = 1, \ldots, p$. Comparison of (3.5) with (3.3) shows that $r_1, \ldots, r_p$ are the inverses of the roots $B_1, \ldots, B_p$ of the polynomial

$$\phi(B) = 0$$

that is, $r_i = B_i^{-1}$. Convergence of γ_k implies, thus, that the roots (in B) of the polynomial $\phi(B)$ are all larger than 1 in modulus. This condition can also be stated as follows: the roots of the polynomial $\phi(B)$ have to lie outside the unit circle (of Figure 2.1(a)). When this happens, we say that the polynomial $\phi(B)$ is stable. From the identity

$$\phi(B)^{-1} = \frac{1}{(1 - r_1 B) \ldots (1 - r_p B)},$$

it is seen that stability of $\phi(B)$ implies, in turn, convergence of its inverse $\phi(B)^{-1}$.

From (2.19), considering that $\Psi(B) = \theta(B)/\phi(B)$, the AGF of z_t is given by

$$\gamma(B, F) = \frac{\theta(B)}{\phi(B)} \frac{\theta(F)}{\phi(F)} V_a, \tag{3.6}$$

and it is straightforward to see that stability of $\phi(B)$ will imply that the stationarity conditions of Section 2.4 are satisfied. The AGF is symmetric and convergent, and the eventual autocorrelation function is the solution of a difference equation, and hence, in general, a mixture of damped polynomials in time and periodic functions. The Fourier transform of (3.6) yields the spectrum of z_t, equal to

$$g_z(\omega) = V_a \frac{\theta(e^{-i\omega})\theta(e^{i\omega})}{\phi(e^{-i\omega})\phi(e^{i\omega})}, \tag{3.7}$$

and the integral of $g_z(\omega)$ over $0 \leq \omega \leq 2\pi$ is equal to $2\pi Var(z_t)$.

A useful result is the following. If two stationary stochastic processes are related through

$$y_t = C(B)x_t,$$

then the AGF of y_t, $\gamma_y(B, F)$, is equal to

$$\gamma_y(B, F) = C(B)C(F)\gamma_x(B, F),$$

where $\gamma_x(B, F)$ is the AGF of x_t. Finally, a function that will prove helpful is the crosscovariance generating function (CGF) between two series, x_t and y_t, with Wold representation

$$\begin{aligned} x_t &= \alpha(B)a_t \\ y_t &= \beta(B)a_t. \end{aligned}$$

Letting $\gamma_j = E(x_t y_{t-j})$ denote the lag-j crosscovariance between x_t and y_t, $j = 0, \pm 1, \pm 2, \ldots$, the CGF is given by

$$CGF(B, F) = \sum_{-\infty}^{\infty} \gamma_j B^j = \alpha(B)\beta(F)\sigma_a^2.$$

If, in equation (3.2), the subindex t is replaced by $t + k$ (k a positive integer), and expectations are taken at time t, the forecast of z_{t+k} made at time t, namely, $\hat{z}_{t+k|t}$, is obtained. Viewed as a function of k (the horizon) and for a fixed origin t, $\hat{z}_{t+k|t}$ is denoted the forecast function. (It is discussed in more detail in Subsection 3.2.3.) Given that $E_t a_{t+k} = 0$ for $k > 0$, it is found that, for $k > q$, the forecast function satisfies the equation

$$\hat{z}_{t+k|t} + \phi_1 \hat{z}_{t+k-1|t} + \ldots + \phi_p \hat{z}_{t+k-p|t} = 0,$$

where $\hat{z}_{t+j|t} = z_{t+j}$ when $j \leq 0$. Therefore, the eventual forecast function is the solution of

$$\phi(B)\hat{z}_{t+k|t} = 0, \tag{3.8}$$

with B operating on k. Comparing (3.4) and (3.8), the link between autocorrelation for lag k (and longer) and k-period-ahead forecast becomes apparent, the forecast being simply an extrapolation of correlation: what we can forecast is the correlation we have detected. For a zero-mean stationary process the forecast function will converge to zero, following, in general, a mixture of damped exponentials and cosine functions.

In summary, stationarity of an ARMA model, which requires the roots (in B) of the autoregressive polynomial $\phi(B)$ to be larger than one in modulus, implies the following model properties: (a) its AGF converges; (b) its forecast function converges; and (c) the polynomial $\phi(B)^{-1}$ converges, so that z_t accepts the convergent (infinite) MA representation

$$z_t = \phi(B)^{-1}\theta(B)a_t = \Psi(B)a_t, \tag{3.9}$$

which is precisely the Wold representation. To see some examples, for the AR(1) model

$$z_t + \phi z_{t-1} = a_t,$$

the root of $1 + \phi B = 0$ is $B_1 = -1/\phi$. Thus stationarity of z_t implies that $|B_1| = |1/\phi| > 1$, or $|\phi| < 1$.

For the AR(2) model

$$z_t + \phi_1 z_{t-1} + \phi_2 z_{t-2} = a_t,$$

stationarity implies that the two roots, B_1 and B_2 are larger than one in modulus. This requires the coefficients ϕ_1 and ϕ_2 to lie inside the triangular region of Figure 3.1. The parabola inside the triangle separates the region associated with complex roots from the one with real roots (Box and Jenkins, 1970, Section 3.2).

If z_t is the differenced series, for which stationarity can be assumed, that is

$$z_t = Dx_t, \qquad D = \nabla^d, \qquad d = 0, 1, 2, \ldots,$$

then the original nonstationary series x_t follows the autoregressive integrated moving average process of orders p, d, and q, or ARIMA(p, d, q) model, given by

$$\phi(B)Dx_t = \theta(B)a_t; \tag{3.10}$$

p and q refer to the orders of the AR and MA polynomials, respectively, and d refers to the number of regular differences (i.e., the number of unit roots at the zero frequency). We often use abbreviated notation, namely,

AR(p): autoregressive process of order p;

MA(q): moving average process of order q;

ARI(p, d): autoregressive process of order p applied to the dth difference of the series;

IMA(d, q): moving-average process of order q applied to the dth difference of the series.

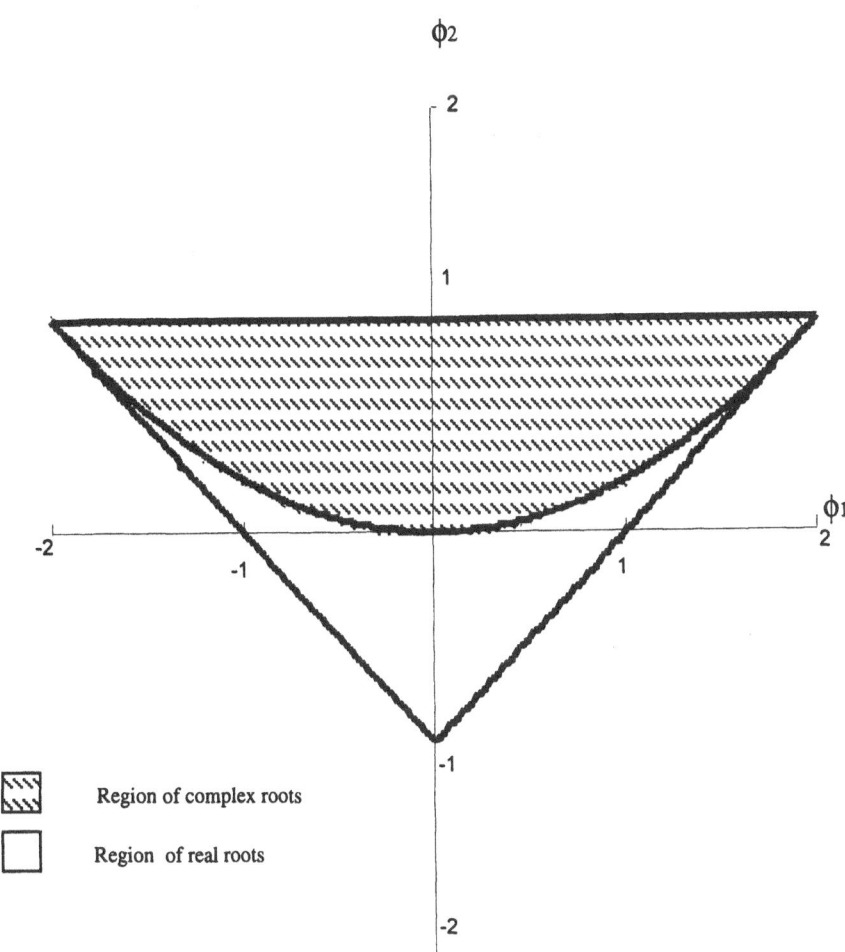

Figure 3.1. Stationary region for AR(2) parameters.

Furthermore, a series is denoted I(d) when it requires d regular differences in order to become stationary.

As in the stationary case, taking conditional expectations at time t in both sides of equation (3.10) with t replaced by $t + k$, where k is a positive integer, it is obtained that

$$\phi(B)D\hat{x}_{t+k|t} = \theta(B)\hat{a}_{t+k|t},$$

where $\hat{x}_{t+j|t} = E(x_{t+j} \mid x_t, x_{t-1}, \ldots)$ is the forecast of x_{t+j} obtained at time t when $j > 0$, and is the observation x_{t+j} when $j \leq 0$; also, $\hat{a}_{t+j|t} = E(a_{t+j} \mid x_t, x_{t-1}, \ldots)$ is equal to 0 when $j > 0$, and is equal to a_{t+j} when $j \leq 0$. As a consequence, the eventual forecast function ($\hat{x}_{t+k|t}$ as a function of k, for $k > q$) will be the solution of the homogeneous difference equation

$$\phi(B)D\hat{x}_{t+k|t} = 0,$$

with B operating on k. The roots of D all have unit modulus; if $D = \nabla^d$polynomial in t of the type $(a + bt^{d-1})$. If D also includes seasonal differencing ∇_4, then the eventual forecast function will also contain the non-convergent deterministic cosine-type function (2.14), associated with the once-and twice-a-year seasonal frequencies, $\omega = \pi/2$ and $\omega = \pi$.

As an example, the forecast function of the model

$$(1 - .7B)\nabla\nabla_4 x_t = (1 + \theta_1 B)(1 + \theta_4 B^4)a_t,$$

consists of five starting values $\hat{x}_{t+j|t}, j = 1, \ldots, 5$, implied by the MA part with $q = 5$, after which the function is the solution of the homogeneous equation associated with the AR part. Factorizing the AR polynomial as

$$(1 - .7B)(1 - B)^2(1 + B)(1 + B^2),$$

the roots of the characteristic equation are given by $r_1 = .7, r_2 = r_3 = 1, r_4 = -1, r_5 = i, r_6 = -i$. From Section 2.3, the eventual forecast function can be expressed as

$$\hat{x}_{t+k|t} = c_1^{(t)}(.7)^k + c_2^{(t)} + c_3^{(t)}k + c_4^{(t)}(-1)^k + c_5^{(t)}\cos\left(\frac{\pi}{2}k + c_6^{(t)}\right),$$

where the last two terms reflect the seasonal harmonics (the root $r_4 = -1$ can also be written as $c_4^{(t)}\cos\pi k$). The constants $c_1, \ldots, c_6$ are determined from the starting conditions of the forecast function, and hence depend on t, the origin of the forecast. This feature gives the ARIMA model its adaptive (or "moving") properties. Notice that, in the nonstationary case, the forecast function (with fixed origin t and increasing horizon k) does not converge.

Concerning the MA polynomial $\theta(B)$, a similar condition of stability is imposed, namely, the roots $B_1, \ldots, B_q$ of the equation $\theta(B) = 0$ have to be larger than one in modulus. This condition is referred to as the invertibility condition for the process and, unless otherwise specified, we assume that the model for the observed series z_t is invertible. This assumption implies that $\theta(B)^{-1}$ converges, so that the model (3.1) can be inverted and expressed as

$$a_t = \theta(B)^{-1}\phi(B)z_t = \Pi(B)z_t, \qquad (3.11)$$

which shows that the series accepts a convergent (infinite) AR expression, and hence can be approximated by a finite AR. Expression (3.11) also shows that, when the process is invertible, the innovations can be recovered from the z_t series.

Some frequency domain implications of nonstationarity and noninvertibility are worth pointing out. Assume that the MA polynomial $\theta(B)$ has a unit root $|B_1| = 1$ –perhaps a complex conjugate pair– associated with the frequency ω_1. Then $\theta(e^{-i\omega_1}) = 0$, and the spectrum of z_t, given by (3.7), will have a zero for the frequency ω_1. Analogously, if $|B_1| = 1$ is a root of the AR polynomial $\phi(B)$, with associated frequency ω_1, then $\phi(e^{-i\omega_1}) = 0$, and $g(\omega_1) \to \infty$.

It follows that

- a unit MA root causes a zero in the spectrum;

- a unit AR root causes a point of ∞ in the spectrum;

- an invertible model will have strictly positive spectrum $g(\omega) > 0$;

- a stationary model has a bounded spectrum $g(\omega) < \infty$.

To illustrate the spectral implications of unit roots, Figure 3.2(a) presents the spectrum of the model

$$(1 - B)x_t = (1 + B)a_t.$$

Since the spectrum is proportional to $(1 + \cos\omega)/(1 - \cos\omega)$, the unit AR root $B = 1$ for the zero frequency makes the vertical axis an asymptote. The unit MA root $B = -1$ for $\omega = \pi$ creates a zero for this frequency. The spectrum of the inverse model

$$(1 + B)x_t = (1 - B)a_t$$

is displayed in Figure 3.2(b). The unit AR root for $\omega = \pi$ implies that the line $\omega = \pi$ is an asymptote, and the unit MA root for $\omega = 0$ implies a spectral zero at the origin.

For quarterly data with seasonality, the differencing D is likely to contain the seasonal difference ∇_4. A popular specification that increases parsimony of the model and permits us to capture seasonal effects is the multiplicative seasonal model

$$\phi(B)\Phi(B^4)\nabla^d\nabla_4^D x_t = \theta(B)\Theta(B^4)a_t, \qquad (3.12)$$

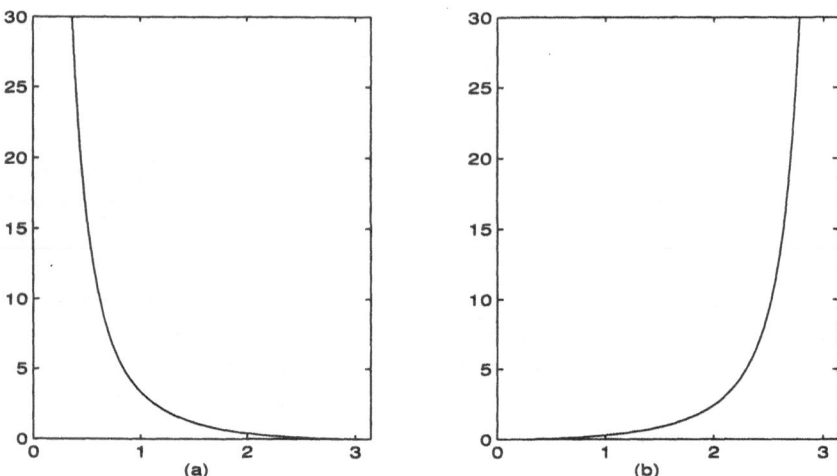

Figure 3.2. Spectra of IMA(1,1) process: (a) spectrum of $\nabla x_t = (1+B)a_t$; (b) spectrum of $\nabla x_t = (1-B)a_t$.

where the regular AR polynomial in B, $\phi(B)$, is as in (3.2), $\Phi(B^4)$ is the seasonal AR polynomial in B^4, d is the degree of regular differencing, D is the degree of seasonal differencing, $\theta(B)$ is the regular MA polynomial in B, $\Theta(B^4)$ is the seasonal MA polynomial in B^4, and a_t denotes the series whitenoise $(0, V_a)$ innovation. The polynomials $\phi(B), \Phi(B^4), \theta(B)$ and $\Theta(B^4)$, are assumed stable, and hence the series

$$z_t = \nabla^d \nabla_4^D x_t$$

follows a stationary and invertible process. (To avoid nonsense complications, we assume that the stationary AR and invertible MA polynomials are prime.) If p, P, q, and Q denote the orders of the polynomials $\phi(B), \Phi(\beta), \theta(B)$, and $\Theta(\beta)$, respectively, where $\beta = B^4$, model (3.12) is referred to as the multiplicative ARIMA $(p, d, q)(P, D, Q)_4$ model. In practice, we can safely restrict the orders to

$$
\begin{array}{ll}
-\quad p, q & \leq 4; \\
-\quad P & \leq 1; \\
-\quad Q & \leq 2; \\
-\quad d & \leq 2; \\
-\quad D & \leq 1.
\end{array}
\qquad (3.13)
$$

Some important practical comments (to bear always in mind) are the following.

1. Parsimony (i.e., few parameters) should be a crucial property of ARIMA models used in practice.

2. ARIMA models are a useful tool for relatively short-term analysis. Their flexibility and adaptive behavior contribute to their good short-term forecasting. Long-term extrapolation of this flexibility may imply, however, unstable long-term inference (see Maravall (1999)). As a general rule, short-term analysis favors differencing, while long-term one favors more deterministic trends that imply less differences.

3.2   Modeling Strategy, Diagnostics and Inference

The so-called Box–Jenkins approach to building ARIMA models consists of the following iterative scheme that contains four stages.

3.2.1   Identification

Two features of the series have to be addressed:

- the degree of regular and seasonal differencing; and

- the orders of the stationary AR and invertible MA polynomials.

Differencing of the series can employ some of the unit root tests available for possibly seasonal data (see, e.g., Hylleberg et al. (1990)). Devised to test deterministic seasonals versus seasonal differencing, these tests are of little use for our purpose. In our experience, stochastic modeling removes in practice the need for the dilemma: deterministic specification versus differencing. Consider, for example, the two models:

(a) $x_t = \mu + a_t$,

(b) $\nabla x_t = (1 - .99B)a_t$.

For a quarterly series, and realistic series length, it is impossible that the sample information can distinguish between the two specifications. Consequently, the choice is arbitrary. Besides the variance of a_t, Model (a) contains one parameter that needs to be estimated, while Model (b) contains none (although, in this case the first observation is lost by differencing). Model (a) offers, thus, no estimation advantage. If short-term forecasting is the main objective, however, Model (b) will display some advantage because it allows for more flexibility, given that it could be rewritten as $x_t = \mu^{(t)} + a_t$, where $\mu^{(t)}$ is a very slowly adapting mean.

A similar consideration applies to seasonal variations. The model

(c) $x_t = \mu + \sum_{j=1}^{3} \beta_j d_{jt} + a_t$,

where d_{jt} denotes a quarterly seasonal dummy variable, is in practice indistinguishable from the direct specification

(d) $\nabla_4 x_t = (1 - .95 B^4) a_t$.

The deterministic specification now has four parameters; the stochastic one has none, but four starting values are lost at the beginning. The latter can also be expressed as

$$x_t = \mu^{(t)} + \sum_{j=1}^{3} \beta_j^{(t)} d_{jt} + a_t,$$

where $\mu^{(t)}, \beta^{(t)}$ denote slowly adapting coefficients. Within our short-term perspective, there is no reason thus to maintain the deterministic-stochastic dichotomy, and deterministic features can be seen as extremely stable stochastic ones.

Besides the lack of power of unit roots tests to distinguish between Models (a) and (b), or (c) and (d), the process of building ARIMA models typically implies estimation of many specifications (if combined with outlier detection and correction, the number may be indeed very large) and the true size of the tests is therefore unknown. In practice, a more efficient and reliable procedure for determining AR unit roots is to use estimation results based on the superconsistency of parameter estimates associated with unit roots, having determined a priori how close to one a root has to be in order to be considered a unit root (see Tiao and Tsay (1983, 1989) and Gómez and Maravall (2000a)).

Once the proper differencing has been established, it remains to determine the orders of the stationary AR and invertible MA polynomials. Here, the basic criterion used to be to try to match the SACF of z_t with the theoretical ACF of a particular ARMA process. In recent years, the efficiency and reliability of automatic identification procedures, based mostly on information criteria, has strongly decreased the importance of the "tentative identification" stage (see Fischer and Planas (1998) and Gómez and Maravall (2000a)).

3.2.2 Estimation and Diagnostics

When $q \neq 0$, the ARIMA residuals are highly nonlinear functions of the model parameters, and hence numerical maximization of the likelihood function, or of some function of the residual sum of squares, can be computationally nontrivial. Within the restrictions in the size of the model given by (3.13), however, maximization is typically well behaved. A standard estimation procedure would cast the model in a state-space format, and use the Kalman filter to compute the likelihood through the prediction error decomposition. The likelihood is then maximized with some nonlinear procedure. Usually, the V_a parameter, as well as a possible constant mean, is concentrated out of the likelihood. When the series is nonstationary, several solutions have been proposed to overcome the problem of defining a proper likelihood. Relevant references are Bell and Hillmer (1991), Brockwell and Davis (1987), De Jong (1991), Gómez and Maravall (1994), Kohn and Ansley (1986), and Morf et al. (1974). Several of these references deal, in fact, with more general models than the straightforward ARIMA one.

Many diagnostics are available for ARIMA models. A crucial one, of course, is the out-of-sample forecast performance. Some tests for in-sample model stability are also of interest. Then, there is a large set of tests based on the model residuals, assumed to be niid. This implies testing for normality, for autocorrelation, for homoscedasticity, and so on. Besides the ones proposed by Box and Jenkins (1970), additional references are Newbold (1983), Gourieroux and Monfort (1990), Harvey (1989), and Hendry (1995).

3.2.3   Inference

If the diagnostics are failed, in the light of the results obtained, the model specification should be changed. When the model passes all diagnostics, we may then proceed to inference. We look in particular at an application in forecasting, unquestionably the main use of ARIMA models.

Let (3.10) denote, in compact notation, the ARIMA model identified for the series x_t, and, as in Section 3.1, denote by $\hat{x}_{t+j|t}$ the forecast of x_{t+j} made at period t (in Box–Jenkins notation, $\hat{x}_{t+j|t} = \hat{x}_t(j)$). Under our assumptions, the optimal forecast of x_{t+j}, in a minimum mean square error (MMSE) sense, is the expectation of x_{t+k} conditional on the observed time series $x_1, \ldots, x_t$ (equal also, to the projection of x_{t+k} onto the observed time series); that is,

$$\hat{x}_{t+j|t} = E(x_{t+k} \mid x_1, \ldots, x_t).$$

This conditional expectation can be obtained with the Kalman filter, or with the Box–Jenkins procedure (for large enough t). Recall that, for known parameters,

$$a_t = x_t - \hat{x}_{t|t-1};$$

that is, the innovations of the process are the sequence of one-period-ahead forecast errors.

The forecast function at time t is $\hat{x}_{t+k|t}$ as a function of k (k a positive integer). In Section 3.1 we saw that for an ARIMA (p, d, q) model, the forecast function consists of q starting conditions, after which it is given by the solution of the homogenous AR difference equation

$$\phi^*(B)\hat{x}_{t+k|t} = 0, \tag{3.14}$$

where B operates on k, and $\phi^*(B)$ denotes the full AR convolution $\phi^*(B) = \phi(B)D$, and includes thus the unit roots.

A useful way to look at forecasts is directly based on the pure MA representation $\Psi(B)$, even in the nonstationary case of a nonconvergent $\Psi(B)$. Assume the model parameters are known and write

$$x_{t+k} = a_{t+k} + \psi_1 a_{t+k-1} + \ldots + \psi_{k-1} a_{t+1} + \psi_k a_t + \psi_{k+1} a_{t-1} + \ldots. \quad (3.15)$$

Given that, for $k > 0$, $E_t a_{t+k} = 0$ and $E_t a_{t-k} = a_{t-k}$, taking conditional expectations in (3.15) yields

$$\hat{x}_{t+k|t} = E_t x_{t+k} = \sum_{j=0}^{\infty} \psi_{k+j} a_{t-j}; \quad (3.16)$$

so that the forecast is a linear combination of past and present innovations. Substracting (3.16) from (3.15), the k-periods-ahead forecast error is given by the model

$$\begin{aligned} e_{t+k|t} &= x_{t+k} - \hat{x}_{t+k|t} \\ &= a_{t+k} + \psi_1 a_{t+k-1} + \ldots + \psi_{k-1} a_{t+1}, \quad (3.17) \end{aligned}$$

an MA(k - 1) process of "future" innovations. From expression (3.17), the joint, marginal, and conditional distributions of forecast errors can be easily derived, and in particular the standard error of the k-period-ahead forecast, equal to

$$SE(k) = (1 + \psi_1^2 + \ldots + \psi_{k-1}^2)^{1/2} \sigma_a. \quad (3.18)$$

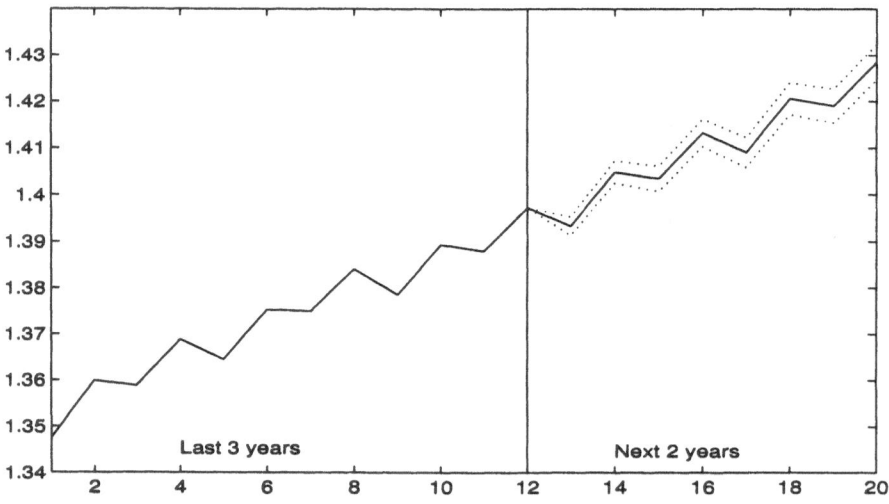

Figure 3.3. Forecasts and 90% confidence intervals.

Unless the series is relatively short, this standard error, estimated by using ML estimators of the parameters, will provide a good approximation. Figure 3.3 displays the last three years of observations and the next two years of ARIMA forecasts for a quarterly series. The forecast function is dominated by a linear trend plus seasonal oscillations; the width of the confidence interval increases with the horizon.

3.2.4   A Particular Class of Models

Box and Jenkins (1970) dedicate a considerable amount of attention to a particular multiplicative model that, for quarterly series, takes the form

$$\nabla\nabla_4 x_t = (1 + \theta_1 B)(1 + \theta_4 B^4)a_t \tag{3.19}$$

(a regular IMA(1,1) structure multiplied by a seasonal IMA(1,1) structure). Given that they identified the model for a series of airline passengers, it has become known as the "airline model". Often, the model is obtained for the logs, in which case a rough first reading shows that the rate of growth of the annual difference is a stationary process.

The model is highly parsimonious, and the three parameters can be given a structural interpretation. As seen in Section 3.1, when $\theta_1 \to -1$, the trend behavior generated by the model becomes more and more stable and, when $\theta_4 \to -1$, the same thing happens to the seasonal component. Estimation of MA roots close to the noninvertibility boundary poses no serious problem, and fixing a priori the maximum value of the modulus of a MA root to, for example, .99 produces perfectly behaved invertible models.

If estimation of (3.19) yields, for example, $\hat{\theta}_4 = -.99$, certain (mutually exclusive) things can explain the result:

(1) seasonality is practically deterministic;

(2) there is no seasonality, and the model is overdifferenced.

Determining which of the two is the correct explanation is rather simple by testing for the significance of seasonal dummy variables. When the model has no seasonality, the seasonal filter $\nabla_4 z_t = (1 - .99B^4)b_t$ would have hardly any effect on the input series. A similar reasoning holds for θ_1 and the possible presence of a deterministic trend. Furthermore, a purely white noise series filtered with model (3.19) with $\theta_1 = \theta_4 = -.99$ would, very approximately, reproduce the series. Thus the airline model also encompasses simpler structures with no trend or no seasonality. Adding the empirical fact that it provides reasonably good fits to many actual macroeconomic series (see, e.g., Fischer and Planas (1998) or Maravall (2000)), it is an excellent model for illustration, for benchmark comparison, and for pretesting.

3.3   Preadjustment

We have introduced the ARIMA model as a practical way of dealing with moving features of series. Still, before considering a series appropriate for ARIMA modeling, several prior corrections or adjustments may be needed. We classify them into the following groups.

1. OUTLIERS

   The series may be subject to abrupt changes that cannot be explained by the underlying normality of the ARIMA model. Three main types of outlier effects are often distinguished: additive outlier, which affects an isolated observation, level shift, which implies a step change in the mean level of the series, and transitory change, similar to an additive outlier whose effect damps out over a few periods. Chen and Liu (1993) suggested an approach to automatic outlier detection and correction that has led to reliable and efficient procedures (see Gómez and Maravall (2000a)).

2. CALENDAR EFFECT

   By this term we refer to the effect of calendar dates, such as the number of working days in a period, the location of Easter effect, or holidays. These effects are typically incorporated into the model through regression variables (see, e.g., Hillmer et al. (1983) and Harvey (1989)).

3. INTERVENTION VARIABLES

   Often special unusual events affect the evolution of the series and cannot be accounted for by the ARIMA model. There is thus a need to "intervene" in the series in order to correct for the effect of special events. Examples can be strikes, devaluations, change of the base index or of the way a series is constructed, natural disasters, political events, important tax changes, or new regulations, to mention a few. These special effects are entered in the model as regression variables (often called, following Box and Tiao (1975), intervention variables).

The full model for the observed series can thus be written as

$$y_t = w_t'\beta + C_t'\eta + \sum_{j=1}^{k} \alpha_j \lambda_j(B) I_t(t_j) + x_t, \qquad (3.20)$$

where $\beta = (\beta_1, \ldots, \beta_n)'$, is a vector of regression coefficients, $w_t' = (w_{1t}, \ldots, w_{nt})$ denotes n regression or intervention variables, C_t' denotes the matrix with columns of the calendar effect variables (trading day, Easter effect, leap year effect, holidays), and η the vector of associated coefficients; $I_t(t_j)$ is an indicator variable for the possible presence of an outlier at period

t_j, $\lambda_j(B)$ captures the transmission of the j-*th* outlier effect (for additive outliers, $\lambda_j(B) = 1$, for level shifts, $\lambda_j(B) = 1/\nabla$, for transitory changes, $\lambda_j(B) = 1/(1 - \delta B)$, with $0 < \delta < 1$), and α_j denotes the coefficient of the outlier in the multiple regression model with k outliers. Finally, x_t follows the general (possibly multiplicative) ARIMA model (3.12). As mentioned earlier, there are several procedures for estimation of models of this type, and easily available programs that enforce the procedures (examples are the programs REGARIMA, see Findley et al. (1998) and TRAMO, see Gómez and Maravall (1996)). Noticing that intervention variables, outliers, and calendar effects are regression variables, the full model can be expressed as a regression-ARIMA model. Estimation typically proceeds by iterating as follows: conditional on the regression parameters (β, η, α), exact maximum likelihood estimation of the ARIMA model is performed; then, conditional on the ARIMA model, GLS estimators of the regression parameters are obtained (both steps can be done with the Kalman filter).

Bearing in mind that preadjustment should be a "must" in applied time series work, for the rest of this book, we assume that the series do not require preadjustment, or have already been subject to one. The series can be directly seen, then, as the outcome of an ARIMA process.

Figures 3.4 and 3.5 illustrate preadjustment in quarterly (simulated) series. The observed original series is displayed in Figure 3.4(a). After removal (through regression) of the outliers automatically identified in the series (two additive outliers, one level shift, and one transitory change) whose effect is displayed in Figure 3.5(a), of the trading-day effect (captured, in this case, with a variable that counts the number of working days) shown in Figure 3.5(b), of the Easter effect, exhibited in Figure 3.5(c), and of an intervention variable associated with the introduction of a regulation that affects the seasonal effect for the last two quarters of each year, the remaining series is displayed in Figure 3.4(b). This is the preadjusted series, also referred to as the "linearized series", given that it can be assumed to be the output of a linear stochastic process (modeled in the ARIMA format).

In the final decomposition of the observed series that we discuss in the following chapters, the different regression effects (outliers, calendar effects, and intervention variables) can be associated with different components. Thus, typically, calendar effects are associated with the seasonal component, additive and transitory outliers are assigned to the irregular component, and level shifts to the trend-cycle component. Care should be taken, however, when a separate business-cycle component is being estimated, because it may require a different allocation of the deterministic effects. For example, when annual data are being used, a transitory change that takes five or six periods to become negligible should probably be included in the cycle, not in the irregular. Likewise, the correction produced by two level shifts of opposite sign and similar magnitude possibly should be assigned to the cycle, not to the long-term trend.

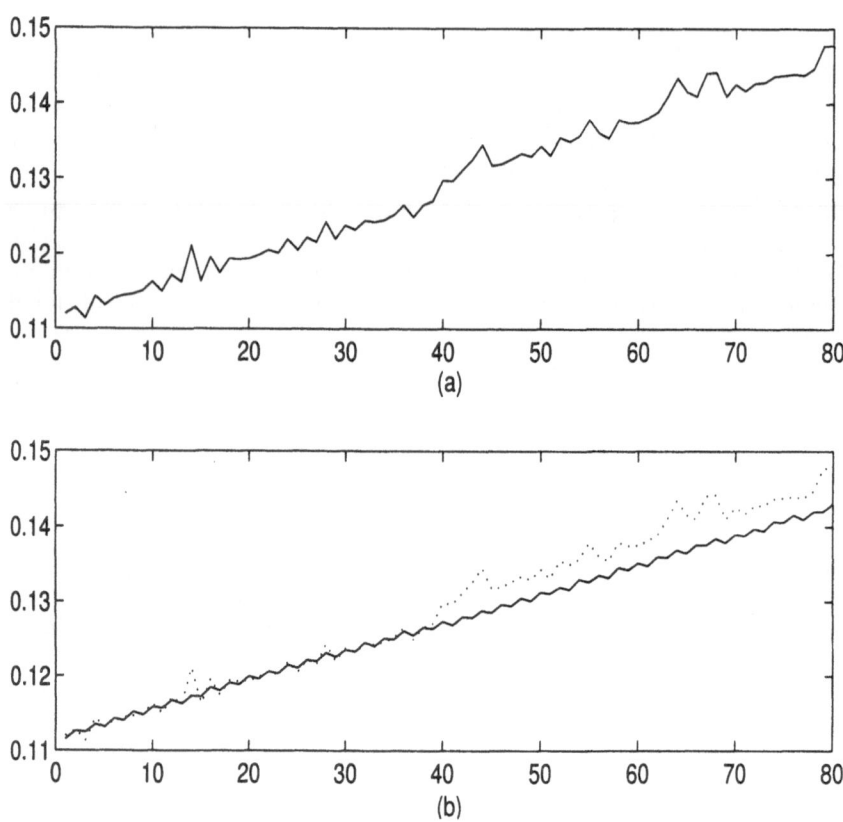

Figure 3.4. Preadjustment: (a) observed series; (b) preadjusted (linearized series).

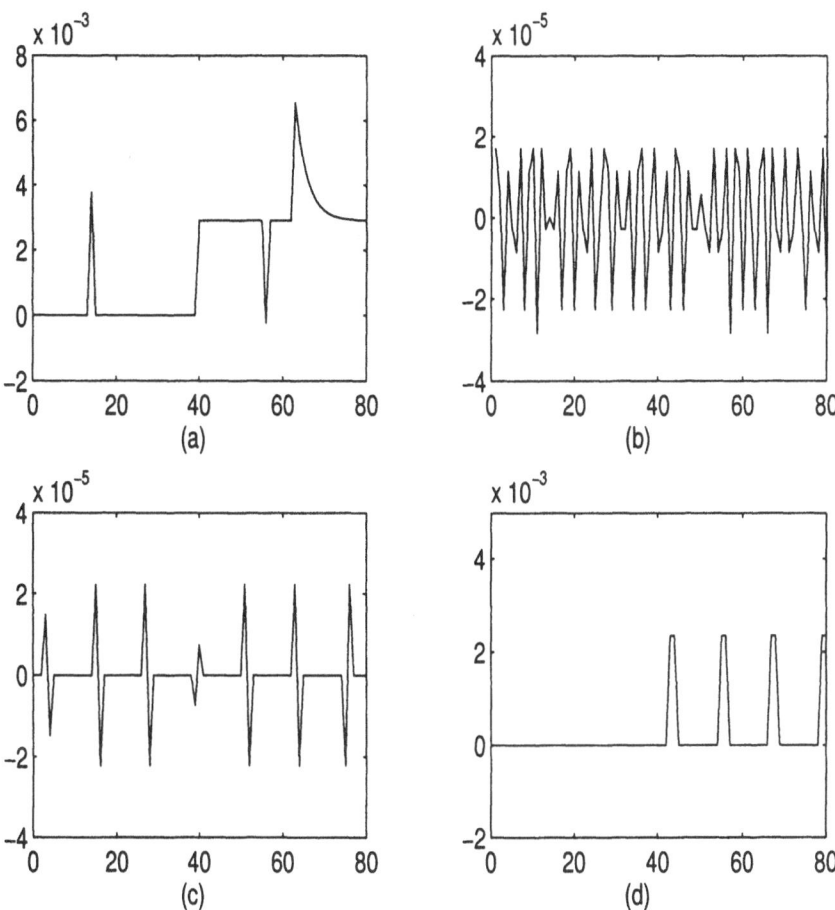

Figure 3.5. Deterministic effects: (a) outlier effect; (b) trading day effect; (c) Easter effect; (d) intervention variable.

3.4   Unobserved Components and Signal Extraction

Assume we are interested in some unobserved component buried in the observed series. Examples can be the seasonally adjusted (SA) series, some underlying short-term trend, or perhaps some cycle. We refer to the component of interest as the signal, and assume it can be extracted from x_t in an additive manner, as in

$$x_t = s_t + n_t, \tag{3.21}$$

where n_t denotes the nonsignal component of the series. (If the signal is the SA series, n_t would be the seasonal component; if the signal is the short-term trend, an additional noise or transitory component may also be included in n_t.) The decomposition can also be multiplicative, as in $x_t = s_t n_t$. Taking logs, however, the additive structure is recovered. For the rest of the discussion we consider the additive decomposition. (A more complete presentation can be found in Planas (1997)).

We further assume that both components are linear stochastic processes, say

$$\phi_s(B)s_t = \theta_s(B)a_{st} \tag{3.22}$$
$$\phi_n(B)n_t = \theta_n(B)a_{nt}. \tag{3.23}$$

The AR polynomials $\phi_s(B)$ and $\phi_n(B)$ also include possible unit roots; in fact, in the vast majority of applications, at least one of the components will be nonstationary. This is because the very concept of a trend or a seasonal component implies a mean that changes with time, and hence a nonstationary behavior that can be removed by differencing.

Concerning expressions (3.22) and (3.23), the following assumptions are made.

(A.1) The variables a_{st} and a_{nt} are mutually independent white noise processes, with zero mean, and variances V_s and V_n, respectively.

(A.2) The polynomials $\phi_s(B)$ and $\phi_n(B)$ are prime.

(A.3) The polynomials $\theta_s(B)$ and $\theta_n(B)$ share no unit root in common.

The first assumption is based on the belief that what causes, for example, seasonality (weather, time of year) is not much related to what may drive a long-term trend (technology, investment), and similarly for other components. Given that different components are associated with different spectral peaks, Assumption A.2 seems perfectly sensible. Assumption A.3 is not strictly needed, but in practice it is hardly restrictive and considerably simplifies notation. The assumption states a sufficient condition for invertibility of the x_t series.

Because aggregation of ARIMA models also yields an ARIMA model, the series x_t follows an ARIMA model, which we write as

$$\phi(B)x_t = \theta(B)a_t, \tag{3.24}$$

where a_t is a white noise variable, $\theta(B)$ is invertible, and $\phi(B)$ is given by

$$\phi(B) = \phi_s(B)\phi_n(B).\tag{3.25}$$

The following identity is implied by (3.22) through (3.24),

$$\theta(B)a_t = \phi_n(B)\theta_s(B)a_{st} + \phi_s(B)\theta_n(B)a_{nt},$$

which shows the relatively complicated relationship between the series innovations and the innovations in the components (see Maravall (1995)).

Having observed a time series $X_T = [x_1, \ldots, x_T]$ our aim is: (1) to obtain minimum mean square error (MMSE) estimators of $\hat{s}_t$ (and $\hat{n}_t$), as well as forecasts; (2) to obtain the full distribution of these estimators, from which diagnostics can be derived; (3) to obtain standard errors for the estimators and forecasts; and (4) to analyze some important features, such as revisions in preliminary estimators, both in terms of size and speed of convergence to the historical estimators.

1. Known models

For the stationary case, the full distribution of (s_t, X_T) is known. Under some additional assumptions (see, e.g., Bell and Hillmer (1991) and Gómez and Maravall (1993)) an appropriate conditional distribution can also be derived for the nonstationary case. The joint distribution is multivariate normal, so that the conditional expectation of the unobserved s_t, given X_T, is a linear combination of the elements in X_T. This conditional expectation also provides the MMSE estimator $\hat{s}_t$, which can thus be expressed as the linear filter

$$\hat{s}_t = E(s_t \mid x_1, \ldots, x_T) = \alpha_1 x_1 + \alpha_2 x_2 + \ldots + \alpha_T x_T.$$

The above conditional expectation can be computed with the Kalman filter (see Harvey (1989) or with the Wiener–Kolmogorov (WK) filter (see Box et al. (1978)). The equivalence of both filters, also when the series is nonstationary, is shown in Gómez (1999). Both filters are efficient; while the Kalman filter has a more flexible format to expand the models, the WK filter is more useful for analysis and interpretation. It is the one used in the discussion.

We start by considering the case of an infinite realization $(x_{-\infty}, \ldots, x_\infty)$. (In practice, this means that we start with historical estimation for the central years of a long-enough series.) As shown in Whittle (1963), the WK filter that yields the MMSE of s_t when model (3.24) is stationary is given by the ratio of the AGF of s_t and x_t, namely,

$$\hat{s}_t = \left[\frac{AGF(s_t)}{AGF(x_t)}\right] x_t = \left[\frac{V_s \dfrac{\theta_s(B)\theta_s(F)}{\phi_s(B)\phi_s(F)}}{V_a \dfrac{\theta(B)\theta(F)}{\phi(B)\phi(F)}}\right] x_t.\tag{3.26}$$

Notice that an important feature of the WK filter (enforced in this way) is that it only requires the specification of the model for the signal, once the model for the observed series has been identified. Contrary to other model-based approaches enforced with the Kalman filter, such as the structural time series model (STSM) approach of Harvey (1989), with the WK filter there is no need to specify the components that aggregate into the non-signal n_t. In view of (3.25), the filter simplifies into

$$\hat{s}_t = \left[k_s \frac{\theta_s(B)\phi_n(B)}{\theta(B)} \frac{\theta_s(F)\phi_n(F)}{\theta(F)} \right] x_t, \qquad (3.27)$$

where $k_s = V_s/V_a$. Direct inspection of (3.27) shows that the filter is the AGF of the stationary model

$$\theta(B)z_t = \theta_s(B)\phi_n(B)b_t, \qquad (3.28)$$

where b_t is white noise with variance (V_s/V_a). The filter is thus convergent in B and F, centered at t, and symmetric.

In order to analyze the properties of the estimated signal, we are interested in its spectrum. If $g_s(\omega), g_n(\omega)$, and $g(\omega)$ denote the spectrum of the signal, the nonsignal component, and the observed series, respectively, orthogonality of s_t and n_t imply

$$g(\omega) = g_s(\omega) + g_n(\omega),$$

where the two components'spectra are nonnegative, and $g(\omega)$ is strictly positive (due to the invertibility condition on the observed series).

The gain of the WK filter, given by the expression in brackets in (3.26), is the Fourier transform of the ratio of two AGFs, so that

$$G(\omega) = g_s(\omega)/g(\omega).$$

Thus, according to (3.26), the spectrum of the MMSE estimator $\hat{s}_t$, denoted $g_{\hat{s}}(\omega)$, is given by

$$
\begin{aligned}
g_{\hat{s}}(\omega) &= \left[\frac{g_s(\omega)}{g(\omega)} \right]^2 g(\omega) \\
&= \left[\frac{g_s(\omega)}{g(\omega)} \right] g_s(\omega) \\
&= G(\omega)g_s(\omega). \qquad (3.29)
\end{aligned}
$$

Given that $G(\omega) \leq 1$, it follows that

$$g_{\hat{s}}(\omega) \leq g_s(\omega),$$

and hence the MMSE estimator will underestimate the variance of the theoretical component.

The filter is well defined everywhere when the ϕ-polynomials contain unit roots, and, in fact, extends, in a straightforward manner, to the nonstationary case (see Bell (1984) and Maravall (1988)). As for the distribution of the estimator $\hat{s}_t$, for the general nonstationary case, assume the polynomial $\phi_s(B)$ can be factorized as

$$\phi_s(B) = \varphi_s(B)D_s,$$

where D_s contains all unit roots, and $\varphi_s(B)$ is a stable polynomial. Multiplying (3.27) by D_s, and replacing $D_s x_t$ by

$$[\theta(B)/\varphi_s(B)\phi_n(B)]a_t,$$

it is obtained that

$$D_s \hat{s}_t = \left[k_s \frac{\theta_s(B)}{\varphi_s(B)} \frac{\theta_s(F)\phi_n(F)}{\theta(F)} \right] a_t, \qquad (3.30)$$

which provides the model that generates the stationary transformation of the estimator $\hat{s}_t$. It is seen that MMSE estimation preserves the differencing of the theoretical component, but has an effect on the stationary structure of the model. The part in B of the model generating the estimator is identical to that of the component; the model for the estimator, however, contains a part in F (that gradually converges towards zero), reflecting the contribution of innovations posterior to t to the historical estimator for period t. The theoretical component, given by (3.22) and the MMSE estimator will have the same stationary transformation, but the AGF and spectra will differ. Furthermore, it is straightforward to see that the AGF of the historical estimation error, $e_t = s_t - \hat{s}_t$, is equal to the AGF of the stationary ARMA model

$$\theta(B)z_t = \theta_s(B)\theta_n(B)b_t, \qquad (3.31)$$

where b_t is white noise with variance $(V_s V_n)/V_a$ (see Pierce (1979)). Stationarity of (3.31) implies that component and estimator are cointegrated.

As was mentioned in Section 2.6, for a finite realization of the x_t process, it will happen that, for periods close enough to both ends of the series, it will not be possible to apply the complete two-sided filter. Denote by $\nu(B, F)$ the filter in brackets in expression (3.27), namely,

$$\nu(B, F) = k_s \frac{\theta_s(B)\phi_n(B)}{\theta(B)} \frac{\theta_s(F)\phi_n(F)}{\theta(F)}, \qquad (3.32)$$

and assume it can be safely truncated after L periods, so that we can write the historical estimator as

$$\hat{s}_t = \nu_0 x_t + \sum_{j=1}^{L} \nu_j(x_{t-j} + x_{t+j}). \qquad (3.33)$$

Let the available time series be $(x_1, \ldots, x_T)$ and, to avoid problems with first observations, let $T > L$. Assume we wish to estimate s_t for $t \leq T$ and $(T - t) > L$, that is, for relatively recent periods. According to (3.33), we need $L - (T - t)$ observations at the end of the filter that are not available yet, namely, $x_{T+1}, x_{T+2}, \ldots, x_{T+L-(T-t)}$. Following Cleveland and Tiao (1986), replacing these future values with the ARIMA forecasts computed at time T, we obtain the preliminary estimator. Rewriting (3.33) as

$$
\begin{aligned}
\hat{s}_t \;=\;& \nu_L x_{t-L} + \ldots + \nu_0 x_t + \ldots + \nu_{(T-t)} x_T \\
+\;& \nu_{(T-t+1)} x_{T+1} + \nu_{(T-t+2)} x_{T+2} \\
+\;& \ldots + \nu_L x_{t+L},
\end{aligned}
\tag{3.34}
$$

and taking conditional expectations at time T, the preliminary estimator of the signal for time t, when observations end at time T, denoted $\hat{s}_{t|T}$, is given by

$$
\begin{aligned}
\hat{s}_{t|T} \;=\;& \nu_L x_{t-L} + \ldots + \nu_0 x_t + \ldots + \nu_{(T-t)} x_T \\
+\;& \nu_{(T-t+1)} \hat{x}_{T+1|T} + \nu_{(T-t+2)} \hat{x}_{T+2|T} \\
+\;& \ldots + \nu_L \hat{x}_{t+L|T},
\end{aligned}
\tag{3.35}
$$

where $\hat{x}_{t_1|t_2}$ denotes the forecasts of x_{t_1} obtained at period t_2. Thus, in compact form, the preliminary estimator can be expressed as

$$
\hat{s}_{t|T} = \nu(B, F) x^e_{t|T},
\tag{3.36}
$$

where $\nu(B, F)$ is the WK filter, and $x^e_{t|T}$ is the "extended" series, such that

$$
\begin{aligned}
x^e_{t|T} \;&=\; x_t && \text{for } t \leq t \\
x^e_{t|T} \;&=\; \hat{x}_{t|T} && \text{for } t > T.
\end{aligned}
$$

The revision the preliminary estimator will undergo until it becomes the historical one is the difference $(\hat{s}_t - \hat{s}_{t|T})$ or, substracting (3.35) from (3.34),

$$
r_{t|T} = \sum_{j=1}^{t+L-T} \nu_{T-t+j} \hat{e}_{T+j|T};
\tag{3.37}
$$

that is, the revision is a linear combination of the forecast errors. Large revisions are unquestionably an undesirable feature of a preliminary estimator, and expression (3.37) shows the close relationship between forecast error and revision: the better we can forecast the observed series, the smaller the revision in the preliminary estimator of the signal will be.

   Direct application of (3.35), when t is close to the end of the series, may require for models close to noninvertibility (for which $\theta(B)^{-1}$ converges slowly) a very large number of forecasts (perhaps more than 100) in order to complete the filter. The Burman–Wilson algorithm (Burman, 1980), described in Section 4.4, permits us to capture, in a very efficient way, the effect of the infinite forecast function with just a small number of forecasts; for the vast majority of quarterly series, 10 forecasts are indeed enough. A similar procedure can be applied to the first periods of the sample to improve starting values for the signal estimator: one can extend the series at the beginning with backcasts (see Box and Jenkins (1970)), and apply the WK filter to the extended series, using a symmetric Burman–Wilson algorithm.

By combining (3.24) with (3.27), an expression is obtained that relates the final estimator $\hat{s}_t$ to the innovations a_t in the observed series, to be represented by

$$\hat{s}_t = \xi_s(B, F)a_t, \tag{3.38}$$

where $\xi_s(B, F)$ can be obtained from the identity

$$\phi_s(B)\theta(F)\xi_s(B, F) = k_s\theta_s(B)\theta_s(F)\phi_n(F), \tag{3.39}$$

and can be seen to be convergent in F. From (3.38), we can write

$$\hat{s}_t = \xi_s^-(B)a_t + \xi_s^+(F)a_{t+1}. \tag{3.40}$$

When t denotes the last observed period, the first term in (3.40) contains the effect of the starting conditions and of the present and past innovations. The second term captures the effect of future innovations (posterior to t). From (3.40), the concurrent estimator is seen to be equal to

$$\hat{s}_{t|t} = E_t s_t = E_t \hat{s}_t = \xi_s^-(B)a_t,$$

so that the revision $r_t = \hat{s}_t - \hat{s}_{t|t}$ is the (convergent) moving average

$$r_t = \xi_s^+(F)a_{t+1}, \tag{3.41}$$

a zero-mean stationary process. Thus historical and preliminary estimators will also be cointegrated. From expression (3.41) it is possible to compute the relative size of the full revision, as well as the speed at which it vanishes.

The distinction between preliminary estimation and forecasting of a signal is, analytically, nonexistent. If we wish to estimate s_t for $t > T$ (i.e., to forecast s_t), expression (3.36) remains unchanged, except that now forecasts start operating "earlier". For example, if the final estimator is given by (3.34) and the concurrent estimator by

$$\hat{s}_{t|t} = \nu_L x_{t-L} + \ldots + \nu_2 x_{t-2} + \nu_1 x_{t-1} + \nu_0 x_t + \sum_{j=1}^{L} \nu_j \hat{x}_{t+j|t},$$

the one-and two-period-ahead forecasts, $\hat{s}_{t|t-1} = E_{t-1}\hat{s}_t$ and $\hat{s}_{t|t-2} = E_{t-2}\hat{s}_t$, will be given by

$$\hat{s}_{t|t-1} = \nu_L x_{t-L} + \ldots + \nu_2 x_{t-2} + \nu_1 x_{t-1} + \sum_{j=0}^{L} \nu_j \hat{x}_{t+j|t-1};$$

$$\hat{s}_{t|t} = \nu_L x_{t-L} + \ldots + \nu_2 x_{t-2} + \sum_{j=-1}^{L} \nu_j \hat{x}_{t+j|t-2} \qquad (\nu_1 = \nu_{-1}),$$

and likewise for other horizons. The discussion on revisions in preliminary estimators applies equally to forecasts. A derivation of the estimation errors associated with the different types of estimators can be found in Maravall and Planas (1999).

2. Unknown models

The previous discussion has assumed known models for the unobserved components s_t and n_t. Given that observations are only available on their sum x_t, quite a bit of a priori information on the components has to be introduced in order to identify and estimate them. Two approaches to the problem have been followed. One, the so-called structural time series model approach, directly specifies models for the components (and ignores the model for the observed series). A trend component p_t will typically follow a model of the type

$$\nabla^d p_t = \theta_p(B) a_{pt}, \qquad (3.42)$$

where $d = 1,2$ and $\theta(B)$ is of order ≤ 2; a seasonal component s_t, will typically be modeled as

$$S s_t = \theta_s(B) a_{st}, \qquad (3.43)$$

with $\theta_s(B)$ also a relatively low-order polynomial in B. Irregular components are often assumed white noise or perhaps some highly transitory ARMA model. As for the cycle, we address the issue in the next chapters. What seems an empirical fact is that direct specification of an ARMA model displaying cyclical behavior due to some complex AR roots with large moduli, as explained earlier, seldom provides significant results.

A limitation of the STSM approach that has often been pointed out is that the a priori structure imposed on the series may not be appropriate for the particular series at hand. This limitation is overcome in the so-called ARIMA model-based approach, where the starting point is the identification of an ARIMA model for the observed series, a relatively well-known problem, and, from that overall model, the appropriate models for the components are derived. These models will be such that their aggregate yields the aggregate model identified for the observations. The models obtained for the trend and seasonal components are also of the type (3.42) and (3.43) and the decomposition may also yield a white noise or a transitory ARMA irregular component. In the applications, we use the program SEATS (signal extraction in ARIMA time series, Gómez and Maravall (1996)). The program originated from the work on AMB decomposition of Burman (1980) and Hillmer and Tiao (1982), done in the context of seasonal adjustment, and proceeded along the lines of Maravall (1995) and Gómez and Maravall (2000b).

Although, as we have presented it, the method can be applied to any signal, it has been developed in the context of the basic "trend-cycle + seasonal component + irregular component" decomposition. A summary of this application will prove helpful.

3.5 ARIMA-Model-Based Decomposition of a Time Series

For the type of quarterly series considered in this work, we briefly summarize the AMB decomposition method. The method starts by identifying an ARIMA model for the observed series. To simplify, assume this model is given by an expression of the type:

$$\nabla \nabla_4 x_t = \theta(B) a_t, \quad a_t \sim \text{niid}(0, V_a), \tag{3.44}$$

where we assume that the model is invertible. Next, components are derived, such that they conform to the basic features of a trend, a seasonal, and an irregular component, and that they aggregate into the observed model (3.44). Considering that $\nabla \nabla_4$ factorizes into $\nabla^2 S$, obviously ∇^2 represents the AR $\phi_p(B)$ polynomial for the trend component, and S represents the AR $\phi_s(B)$ polynomial for the seasonal component. The series is seen to contain a nonstationary trend (or trend-cycle) and seasonal components, and it can be decomposed into

$$x_t = p_t + s_t + u_t, \tag{3.45}$$

where p_t, s_t, and u_t denote the trend-cycle, seasonal, and irregular components, respectively, the latter being a stationary process. When q (the order of $\theta(B)$) ≤ 5, the following models for the components are obtained

$$\nabla^2 p_t = \theta_p(B) a_{pt}, \quad a_{pt} \sim \text{niid}(0, V_p)$$
$$S s_t = \theta_s(B) a_{st}, \quad a_{st} \sim \text{niid}(0, V_s) \tag{3.46}$$
$$u_t \sim \text{niid}(0, V_u),$$

where a_{pt}, a_{st}, and u_t are mutually uncorrelated white noise variables. We refer to (3.46) as the (unobserved component) "structural model" associated with the reduced form model (3.44). Applying the operator $\nabla \nabla_4$ to both sides of (3.45), the identity

$$\theta(B) a_t = S \theta_p(B) a_{pt} + \nabla^2 \theta_s(B) a_{st} + \nabla \nabla_4 u_t \tag{3.47}$$

is obtained. If the lhs. of (3.47) is an MA(5) process, setting the order of $\theta_p(B)$, q_p, equal to 2, and that of $\theta_s(B)$, q_s, equal to 3, all terms of the sum in the rhs of (3.47) are also MA(5)s. Thus we assume, in general, $q_p = 2, q_s = 3$, and equating the AGF of both sides of (3.47), a system of six equations is obtained (one equation for each nonzero covariance). The unknowns in the system are the two parameters in $\theta_p(B)$, the three parameters in $\theta_s(B)$, plus the variances V_p, V_s, and V_u; a total of eight unknowns. There are not enough equations to identify the parameters, and hence there is, as a consequence, an infinite number of solutions to (3.47).

Denote a solution that implies components as in (3.46) with nonnegative spectra an admissible decomposition. The structural model will not be identified, in general, because an infinite number of admissible decompositions are possible. The AMB method solves this underidentification problem by maximizing the variance of the noise V_u, which implies inducing a zero in the spectra of p_t and s_t in (3.46). The spectral zero translates into a unit root in $\theta_p(B)$ and in $\theta_s(B)$, so that the two components p_t and s_t become noninvertible.

This particular solution to the identification problem is referred to as the "canonical" decomposition (see Box et al. (1978) and Pierce (1978)); from all infinite solutions of the type (3.46), the canonical one maximizes the stability of the trend-cycle and seasonal components that are compatible with the model (3.44) for the observed series. Furthermore, the trend-cycle and seasonal components for any other admissible decomposition can be expressed as the canonical ones perturbated by orthogonal white noise. Also, if the model accepts an admissible decomposition, then it accepts the canonical one (see Hillmer and Tiao (1982)). Notice that, since it should be a decreasing function of ω in the interval $(0, \pi)$, the spectrum of p_t should display the zero at the frequency π. Thus the trend-cycle MA polynomial can be factorized as $\theta_p(B) = (1+\alpha B)(1+B)$, where the root $B = -1$ reflects the spectral zero at π (see Section 2.5). The zero in the spectrum of s_t may occur at $\omega = 0$ or at a frequency roughly halfway between the two seasonal frequencies $\omega = \pi/2$ and $\omega = \pi$.

One simple example may clarify the canonical property. Assume an unobserved component model for which the trend follows the random walk model

$$\nabla p_t = a_{pt}, \qquad V_p = 1.$$

This specification is in fact often found in macroeconomic applications of unobserved component models (Stock and Watson, 1988). Part (a) of Figure 3.6 displays the spectrum of p_t. It is clear that it does not satisfy the canonical condition because

$$\min_\omega g_p(\omega) = g_p(\pi) = .25 > 0.$$

It is straightforward to check that the trend p_t can be decomposed into a canonical trend p_t^*, plus orthogonal white noise u_t, according to

$$p_t = p_t^* + u_t,$$

where

$$\nabla p_t^* = (1 + B)a_{pt}^*,$$

with $V_{p^*} = .25$. Part (b) of Figure 3.6 shows the spectral decomposition of the random walk. The canonical p_t^* is clearly smoother, since it has removed white noise from p_t. The spectral zero for $\omega = \pi$ of the canonical trend is associated with the $(1 + B)$ MA polynomial, with unit root $B = -1$.

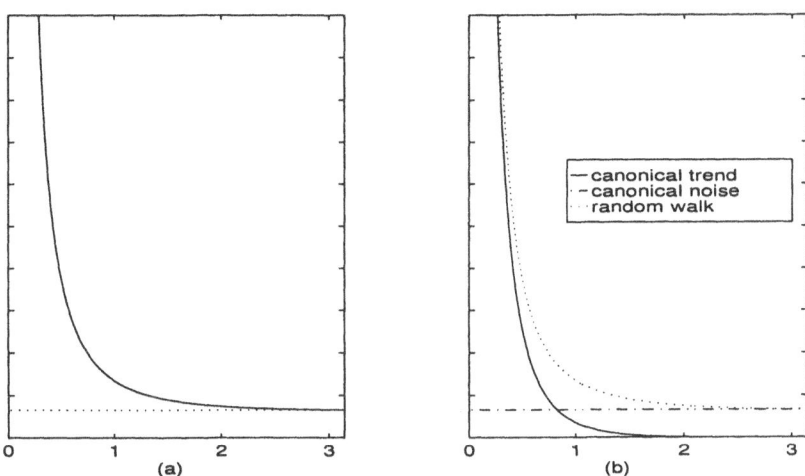

Figure 3.6. Canonical decomposition of a random walk: (a) random walk; (b) canonical decomposition.

One relevant property of noninvertible series (and hence, of canonical components) is that, due to the spectral zero, no further noise can be extracted from them.

The AMB method computes the trend-cycle, seasonal, and irregular component estimators as the MMSE ("optimal") estimators based on the available series $X_t = [x_1, \ldots, x_T]$, as described in the previous section. Under our assumptions, these estimators are also conditional expectations of the type $E(\text{component} \mid [\text{observed series}])$, and they are obtained using the WK filter. For a series extending from $t = -\infty$ to $t = \infty$, that follows model (3.44), assume we are interested in estimating a component, which we refer to as the "signal" (the signal will be p_t, then s_t, and finally u_t). Applying result (3.28) to the model (3.46), the WK filter for historical estimation of the trend-cycle component is equal to the AGF of the model

$$\theta(B)z_t = [\theta_p(B)S]b_t, \quad b_t \sim \text{niid}(0, V_p/V_a); \tag{3.48}$$

for the seasonal component it is given by the AGF of

$$\theta(B)z_t = [\theta_s(B)\nabla^2]b_t, \quad b_t \sim \text{niid}(0, V_s/V_a); \tag{3.49}$$

and for the irregular component, by the AGF of

$$\theta(B)z_t = \nabla\nabla_4 b_t, \quad b_t \sim \text{niid}(0, V_u/V_a). \tag{3.50}$$

Notice that this last model is the "inverse" model of (3.44), which is assumed known. Also invertibility of (3.44) guarantees stationarity of the models in (3.48) through (3.50), and hence the three WK filters will converge in B and in F.

For a finite realization, as already mentioned, the optimal estimator of the signal is equal to the WK filter applied to the available series extended with optimal forecasts and backcasts, obtained with (3.44). This is done with the Burman-Wilson algorithm referred to in the previous section.

The following figures illustrate the procedure. Figure 3.7 shows the spectrum of a particular case of model (3.44) and its spectral decomposition into trend, seasonal, and irregular components. The trend captures the peak around $\omega = 0$, and the seasonal component the peaks around the seasonal frequencies. Figure 3.8 displays the WK filters to obtain the historical estimates of the SA series, trend, seasonal, and irregular components. From Figures 3.8(a) and (b), it is seen, for example, that the concurrent estimator of the SA series requires many more periods to converge to the historical one than that of the trend. Figure 3.9 shows the squared gains of the WK filter (see Section 2.6), that is, which part of the series variation is passed to, or cut off from, each component. As seen in Figure 3.9(c), to estimate the irregular component only the frequencies of no interest for the trend or seasonal component are employed. Figure 3.10(a) exhibits a time series of 100 observations generated with the model of Figure 3.7(a), and Figures 3.10(b),(c), and (d) the estimates $\hat{n}_{t|100}, \hat{p}_{t|100}$, and $\hat{s}_{t|100}(t = 1, \ldots, 100)$ of the trend, seasonal, and irregular components. Figure 3.11 presents the standard errors of the estimates of Figure 3.10, moving from concurrent to final estimator. The trend estimator converges in a year, while the SA series takes about three years for convergence. Finally, Figure 3.12 presents the forecast function of the original series, trend and seasonal components, as well as the associated 90% probability intervals.

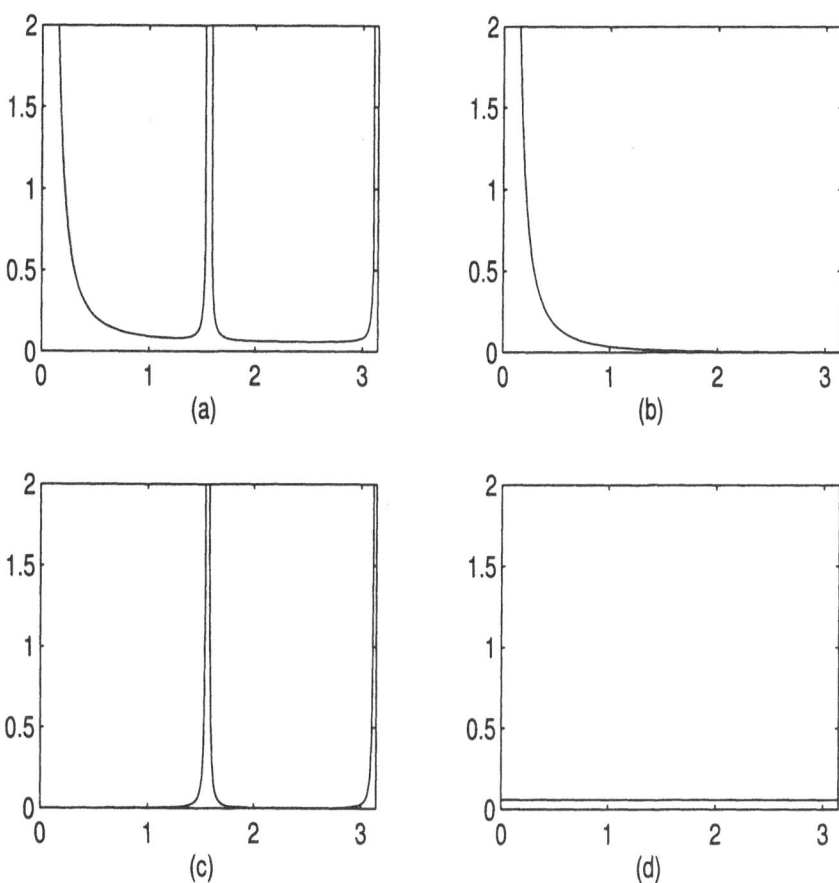

Figure 3.7. Spectral AMB decomposition: (a) spectrum series; (b) spectrum trend; (c) spectrum seasonal; (d) spectrum irregular.

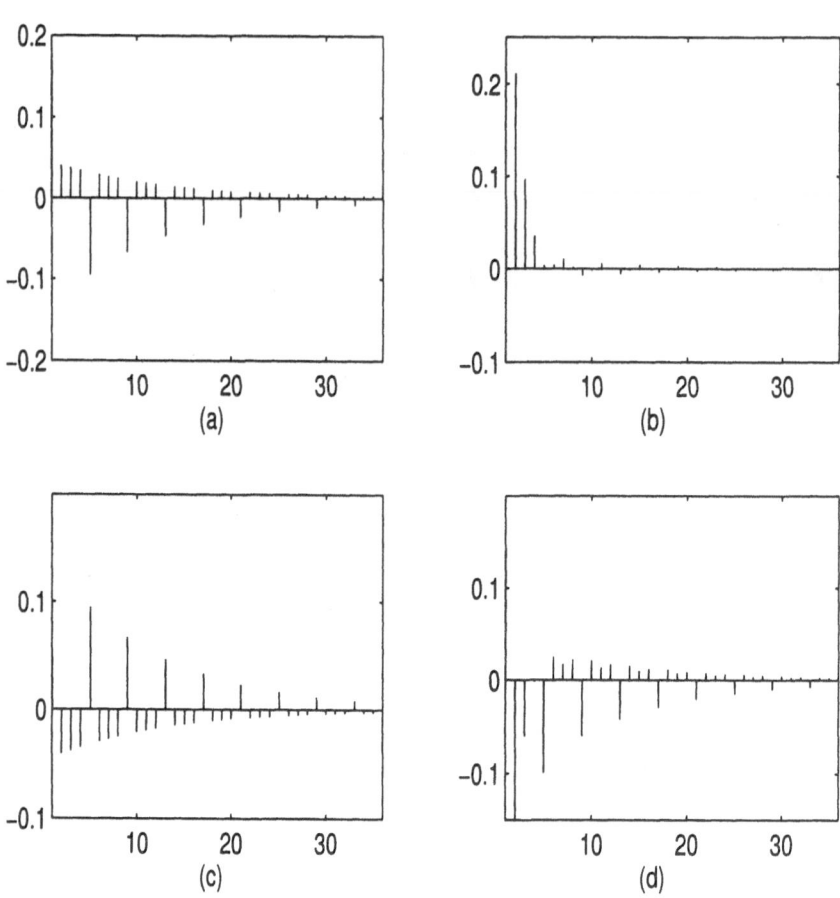

Figure 3.8. Wiener–Kolmogorv filters: (a) filter for SA series; (b) filter for trend; (c) filter for seasonal component; (d) filter for irregular component.

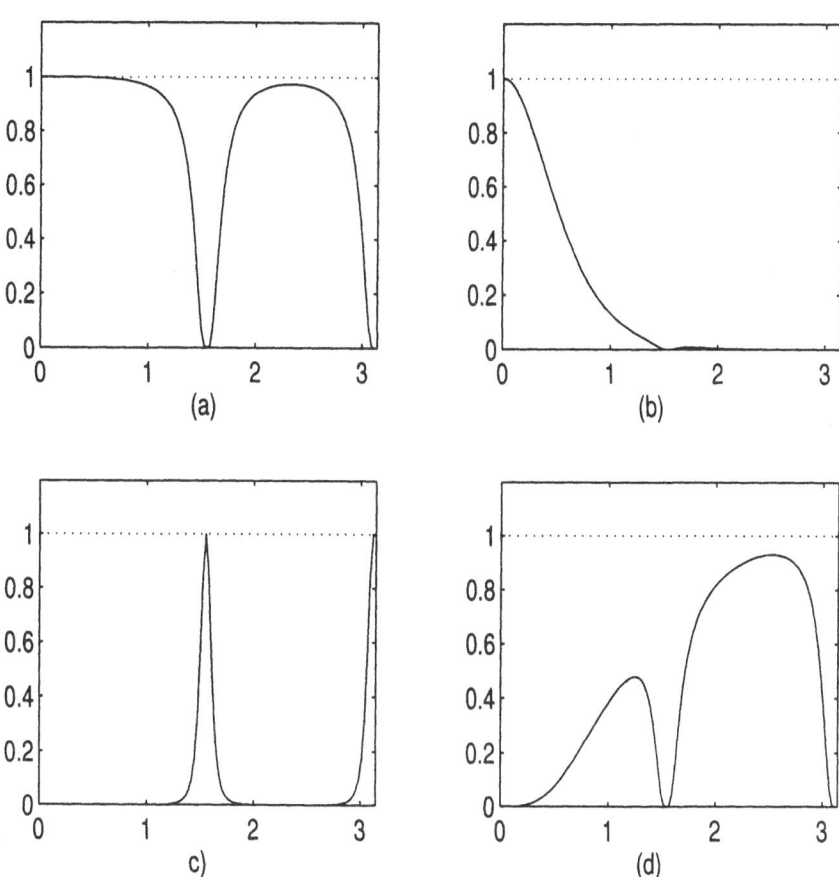

Figure 3.9. Squared gains: (a) SA filter; (b) trend-cycle component filter; (c) seasonal component filter; (d) irregular component filter.

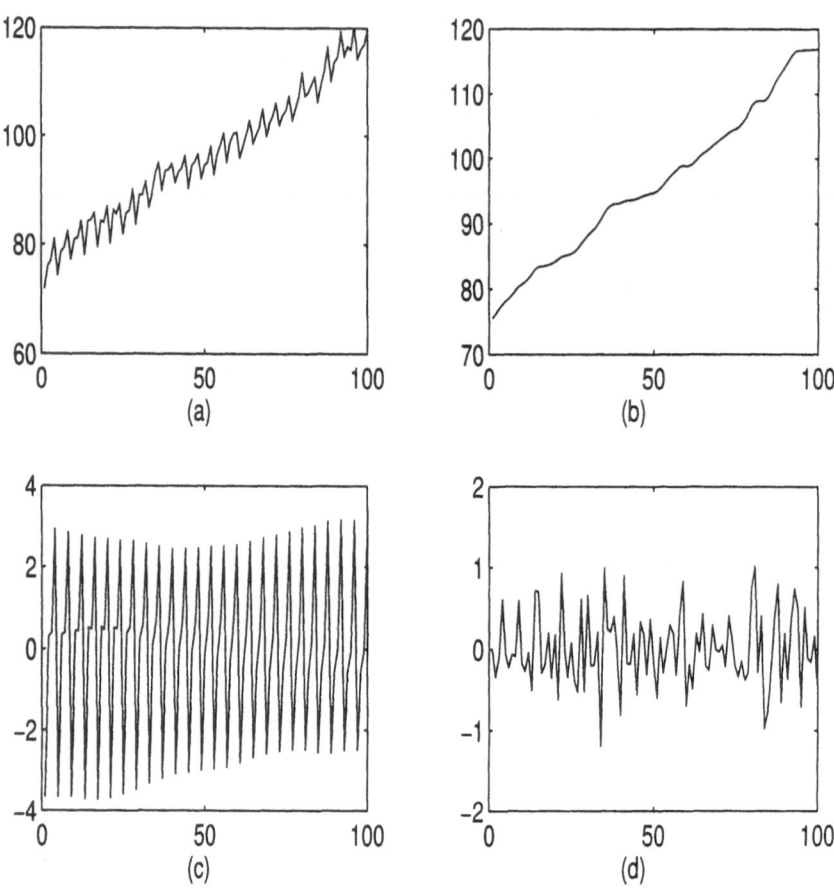

Figure 3.10. Series and estimated components: (a) original series; (b) trend-cycle component; (c) seasonal component; (d) irregular component.

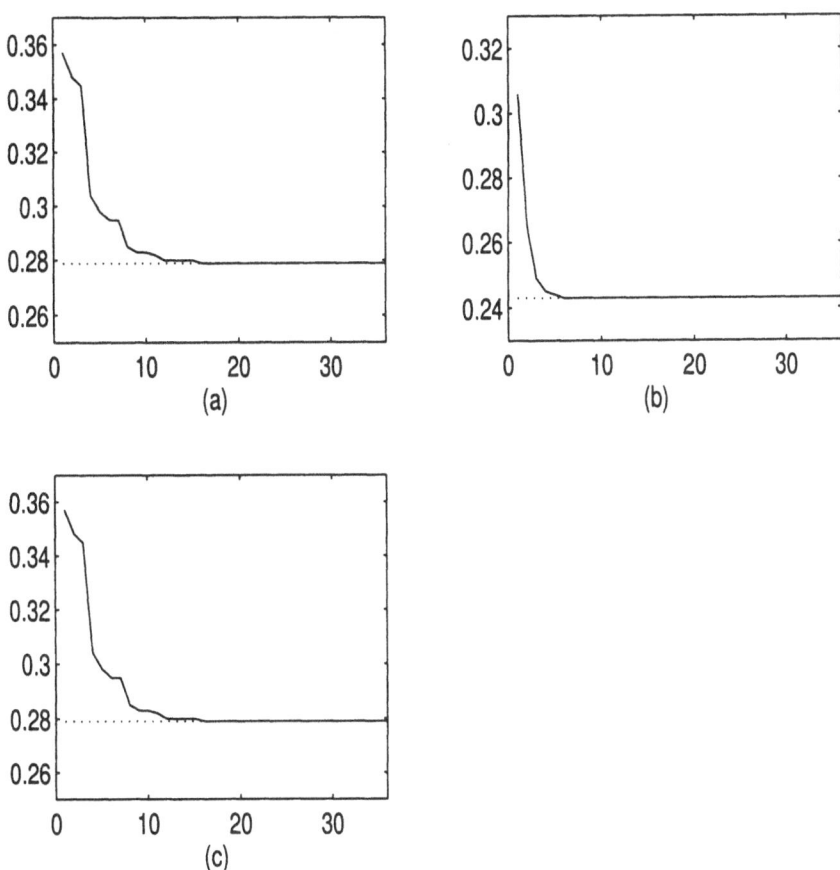

Figure 3.11. Standard error of estimators: (a) SA series; (b) trend-cycle component; (c) seasonal component.

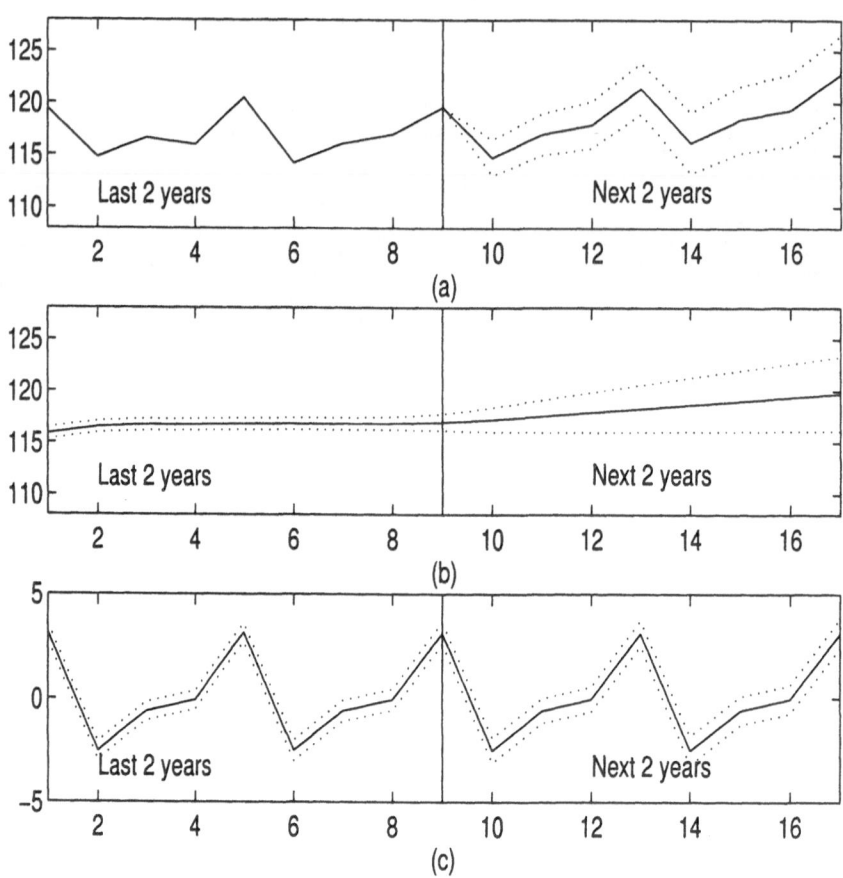

Figure 3.12. Forecasts: (a) original series; (b) trend-cycle component; (c) seasonal component.

3.6   Short-Term and Long-Term Trends

The previous figures serve also to illustrate an important point, often a source of confusion, namely, the meaning of a trend component. It is a well-known fact that the width of the spectral peak for $\omega = 0$ in parsimonious ARIMA models may vary considerably, so that the same will be true for the squared gain of the trend estimator. Figure 3.13 shows these squared gains for model (3.19), for different combinations of the θ_1 and θ_4 parameters. If the range of cyclical frequencies is broadly defined as starting slightly to the right of $\omega = 0$, and finishing slightly to the left of the fundamental frequency ($\omega = \pi/2$) (so that cycles have periods longer than a year, yet reasonably bounded), then Figure 3.13 shows how the squared gain of the trend filter may very well extend over the range of cyclical frequencies, and even exhibit spill-over effects for higher frequencies. This feature is also typical of the squared gains derived from the structural time series model approach (see Harvey (1989) and Koopman et al. (1996)), and from well-known detrending filters such as the Henderson ones implemented in the X11 family of programs (see Findley et al. (1998)).

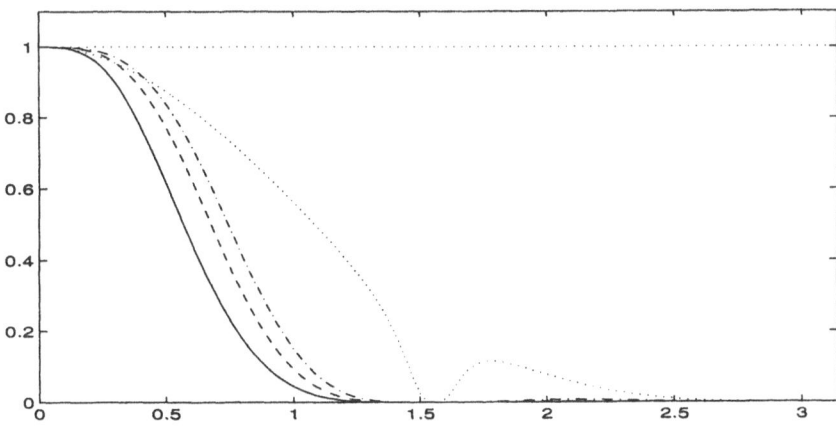

Figure 3.13. Squared gain for trend-cycle filter: different ARIMA models.

As a consequence, the trend estimators obtained with these procedures may contain a large amount of relatively short-term variation. These short-term trend components should be more properly called trend-cycle components. The contamination of trend with cyclical frequencies is clearly a result of the implicit definition of the trend in the decomposition (3.44). The two components that are removed from the series in order to obtain the trend are the seasonal component and the highly transitory (close to white) noise component. Therefore, the trend is basically defined as the "noise-free SA series", and includes, as a consequence, cyclical frequencies.

Its interest lies in the belief that noise, unrelated to the past and to the future, is more disturbing than helpful in short-term monitoring of the series. (In fact, SA series and trend-cycle components for short-term indicators are both provided at several data-producing agencies; see Eurostat (1999) and Banco de España (1999).) A discussion of short-term trends is contained in Maravall (1993).

Another important area where trends are used is business-cycle analysis. Here, the trend is also defined as the detrended and SA series, but the concept of detrending is rather different. The aim is to remove a long-term trend that does not include movements with periods shorter than a certain number of years (roughly, the cutting point is set within the range 8 to 10 years). Having defined the band in the frequency range associated with cyclical oscillations (e.g., those with period between 2 and 10 years), the issue is to design a band-pass filter that permits only the passage of frequencies within that band. Linear filters can only do this job in an approximate manner because the first derivative with respect to ω of their squared gain function is everywhere well defined, and cannot take the form of an exact rectangle, with base the frequency band pass, and height one. The Butterworth family of filters was designed to approximate this band-pass feature. One of the members of the family is very well-known in economics, where it is usually called the Hodrick–Prescott filter (see Hodrick and Prescott (1980) or Prescott (1986)). Despite the fact that business cycle estimation is basic to the conduct of macroeconomic policy and to monitoring of the economy, many decades of effort have shown that formal modeling of economic cycles is a frustrating issue. As a consequence, applied work and research at economic-policy related institutions has relied (and still relies) heavily on ad hoc band-pass filters and, in particular, on the HP one. One can say that HP filtering of X11-SA series has become the present paradigm for business-cycle estimation in applied work. Figure 3.14 represents, for the example of the previous section, the short-term trend (or trend-cycle component) obtained with the AMB approach, and the long-term trend obtained with the HP-X11 filter. Part (a) compares the two squared gains, and part (b) the two estimated trends. The short-term character of the AMB trend and the long-term character of the X11-HP trend are clearly discernible.

If business-cycle analysts complain that series detrended with short-term trends, of the type obtained in the AMB approach, contain very little cyclical information, ad hoc fixed filters to estimate long-term trends are criticized because the trends they yield could be spurious. As we show, however, the two types of trends are not in contradiction and can be instead quite complementary. When properly used, their mixture can incorporate the desirable features of the ad hoc design, with a sensible and complete model-based structure, that fully respects the features of the series at hand. For the rest of the book, short-term trends are denoted trend-cycle components and are represented by p_t; long-term trends are simply called trends, and represented by m_t.

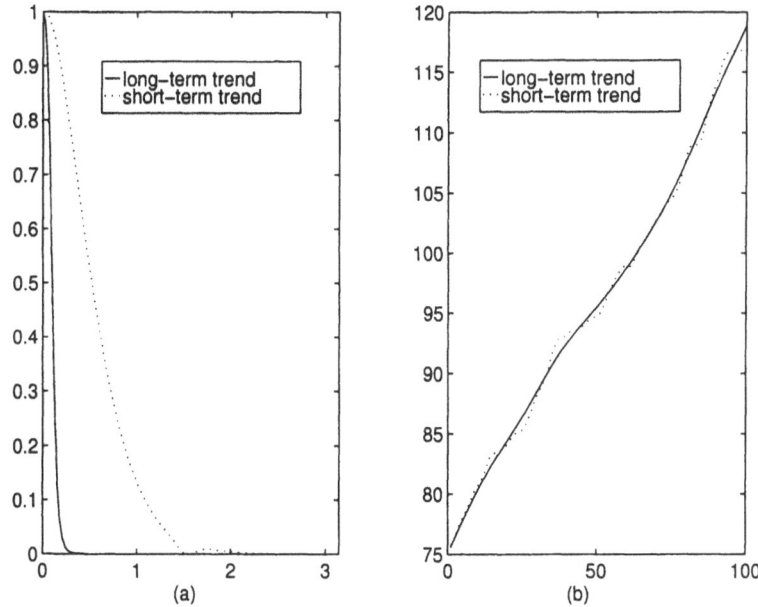

Figure 3.14. (a) Squared gains of trend filters; (b) trend estimator.

4.

Detrending and the Hodrick–Prescott Filter

4.1 The Hodrick–Prescott Filter: Equivalent Representations

A well-known family of ad hoc filters, designed to capture a band of frequencies in the low frequency range, is the Butterworth family of filters (see, e.g., Otnes and Enochson (1978)). The filters are typically expressed by means of the sine or tangent functions. For the sine-type case, the filter is represented, in the frequency domain, by its gain function which, for the two-sided filter, is given by

$$G(\omega) = \left[1 + \left(\frac{\sin(\omega/2)}{\sin(\omega_0/2)}\right)^{2d}\right]^{-1} ; \quad 0 \le \omega \le \pi. \tag{4.1}$$

The filter depends on two parameters that have to be specified a priori: the frequency ω_0 defined as the frequency for which the gain is .5 (hence $G(\omega_0) = .5$); and the parameter d, a positive integer. To obtain the time domain equivalent of (4.1), we use the trigonometric identity

$$4\sin^2(\omega/2) = (1 - e^{-i\omega})(1 - e^{i\omega}),$$

and set the complex variable $B = e^{-i\omega}$. This yields

$$\nu(B, F) = \frac{1}{1 + \lambda[(1 - B)(1 - F)]^d}, \tag{4.2}$$

where

$$\lambda = [4\sin^2(\omega_0/2)]^{-d} \tag{4.3}$$

is a constant. Rewriting (4.2) as

$$\nu(B, F) = \frac{\dfrac{1}{\lambda[(1 - B)(1 - F)]^d}}{1 + \dfrac{1}{\lambda[(1 - B)(1 - F)]^d}},$$

direct comparison with (3.26) shows that $\nu(B, F)$ is the WK filter to estimate the signal s_t in the "signal plus noise" decomposition of a series x_t, where the signal is the IMA(d,0) model

$$\nabla^d s_t = a_{st},$$

the noise is white, and $\lambda = Var(\text{noise})/Var(a_{st})$. When $d = 1$, (4.1) or (4.2) would decompose the series into a random walk plus white noise. When $d = 2$, the signal is given by the model $\nabla^2 s_t = a_{st}$. Figure 4.1 compares the squared gain of different Butterworth filters. Smaller values of ω_0 narrow the width of the filter band; higher values of d make the slope of the band more vertical.

Ad hoc filters have also been constructed using an alternative approach called the penalty-function method. In the decomposition of x_t into $s_t + n_t$, this approach obtains the filter to estimate the signal s_t as the solution to a problem that balances a trade-off between fit (small values of the noise n_t) and smoothness of the signal. In particular, this approach provides the standard derivation of the Hodrick–Prescott filter. Changing the notation "signal plus noise" by "trend plus cycle", let x_t ($t = 1, \ldots, T$) denote an observed series. The HP filter decomposes x_t into a smooth trend (m_t) and a residual (c_t), where the trend is meant to capture the long-term growth of the series, and the residual (equal to the deviation from that growth) represents the cyclical component.

Since seasonality should not contaminate the cycle, the filter is typically applied to SA series, but for the moment we assume that the series contains no seasonality. In the decomposition

$$x_t = m_t + c_t, \tag{4.4}$$

the HP filter provides the estimator of c_t and m_t such that the expression

$$\sum_{t=1}^{T} c_t^2 + \lambda \sum_{t=3}^{T} (\nabla^2 m_t)^2 \tag{4.5}$$

is minimized. The first summation in (4.5) penalizes bad fitting, while the second one penalizes lack of smoothness (the smoother m_t is, the smaller will be —in absolute value— its second differences). The parameter $= 0$, $\hat{m}_t = x_t$, when $\lambda \to \infty$, $\hat{m}_t$ becomes a deterministic linear trend.

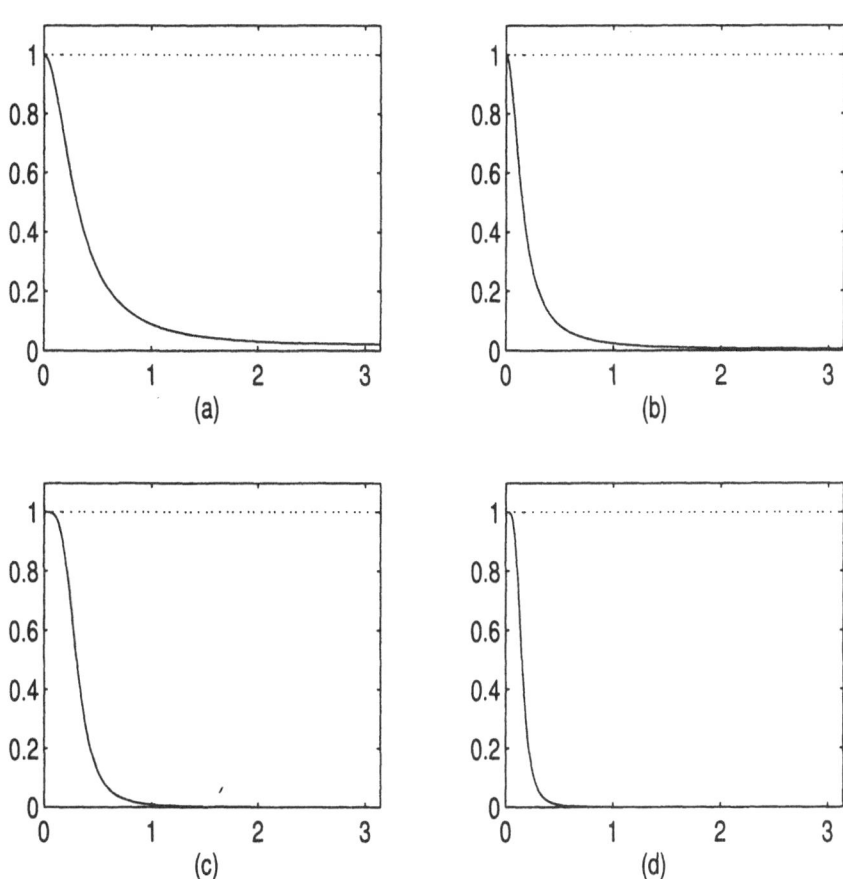

Figure 4.1. Squared gain function: Butterworth filters. (a) $\omega_0 = .30, d = 1$; (b) $\omega_0 = .15, d = 1$; (c) $\omega_0 = .30, d = 2$; (d) $\omega_0 = .15, d = 2$.

The solution to the problem of minimizing (4.5) subject to the restriction (4.4) is given by (see Danthine and Girardin (1989))

$$\hat{\mathbf{m}} = \mathbf{A}^{-1}\mathbf{x}, \qquad \mathbf{A} = \mathbf{I} + \lambda \mathbf{K}'\mathbf{K}, \tag{4.6}$$

where $\hat{\mathbf{m}}$ and $\mathbf{x}$ are the vectors $(\hat{m}_1, \ldots, \hat{m}_T)'$ and $(x_1, \ldots, x_T)'$, respectively, and $\mathbf{K}$ is an $(T-2) \times T$ matrix with its elements given by $K_{ij} = 1$ if $i = j$ or $i = j + 2$, $K_{ij} = -2$ if $i = j + 1$, and $K_{ij} = 0$ otherwise.

Clearly, the estimator of the trend for a given period will depend on the length of the series. Consider the trend for period T, the last observed period. Application of (4.6) yields an estimator that is denoted $\hat{m}_{T|T}$, where the first subindex refers to the period under estimation, and the second to the last observed period. This estimator is called the concurrent estimator. When one more quarter is observed and $\mathbf{x}$ becomes $(x_1, \ldots, x_{T+1})'$, application of (4.6) yields a new estimation of m_T, namely, $\hat{m}_{T|T+1}$. As more quarters are added, the estimator is revised. It is easily seen that, for large enough k, $\hat{m}_{T|T+k}$ converges to a final or historical estimator, denoted $\hat{m}_T$. Therefore, for a long enough series, the final estimator may be assumed for the central periods, while estimators for the last years are preliminary. This two-sided interpretation of the HP filter seems unavoidable. Because additional correlated new information cannot deteriorate an estimator, $\hat{m}_{T|T+1}$ should improve upon $\hat{m}_{T|T}$. Moreover, actual behavior of the US Business Cycle Dating Committee (or similar institutions) reveals in fact a two-sided filter, which starts with a preliminary estimator, and reaches the final decision with a lag of perhaps two years.

As shown in King and Rebelo (1993), the HP filter can be given a model-based interpretation. Let c_t in (4.4) be white noise with variance V_c, and let m_t follow the IMA(2,0) model

$$\nabla^2 m_t = b_t, \tag{4.7}$$

where b_t is a white noise variable (with variance V_m) uncorrelated to c_t. From the results in Section 3.5, the filter (4.2) provides the MMSE estimator of m_t in the King and Rebelo interpretation, and hence the HP filter can also be seen as a Butterworth-type filter of the sine type, with $d = 2$ and $\lambda = V_c/V_m$. (Without loss of generality, we can set $V_c = \lambda$, $V_m = 1$.) The Kalman filter provides a convenient algorithm to obtain the MMSE estimator of m_t in the decomposition (4.4), with m_t given by (4.7) and c_t white noise (see Harvey and Jaeger (1993)).

4.2   Basic Characteristics of the Hodrick–Prescott Filter

The filter depends on one parameter, λ, that needs to be set a priori. Its interpretation varies according to the filter derivation. In the penalty-function derivation, λ regulates the trade-off between fit and smoothness in the function (4.5) that is to be minimized. When derived as the MMSE estimator of an IMA(2,0) trend (4.7) buried in white noise, λ is equal to the ratio of the noise and trend innovation variances. Finally, when derived as a Butterworth filter, it is related to the frequency ω_0 (for which $G(\omega_0) = 1/2$) through equation (4.3); that is, λ is the parameter such that 50% of the filter gain has been completed for the frequency

$$\omega_0 = 2\arcsin\left(\frac{1}{2\lambda^{1/4}}\right); \qquad 0 \le \omega_0 \le \pi; \qquad \lambda > 0. \qquad (4.8)$$

Figure 4.2a compares the squared gain of the HP trend-filter for the values $\lambda = 400$, 1600, and 6400. The associated frequencies are $\omega_0 = .3176, .1583,$ and .1119 so that, from (2.13), when $\lambda = 400$, 50% of the filter gain is achieved for a period of 20 quarters, when $\lambda = 1600$, for a period of 40 quarters, and when $\lambda = 6400$, for a period of 56 quarters. Thus, larger values of λ narrow the width of the trend filter gain, and hence provide smoother trends, of a longer-term nature.

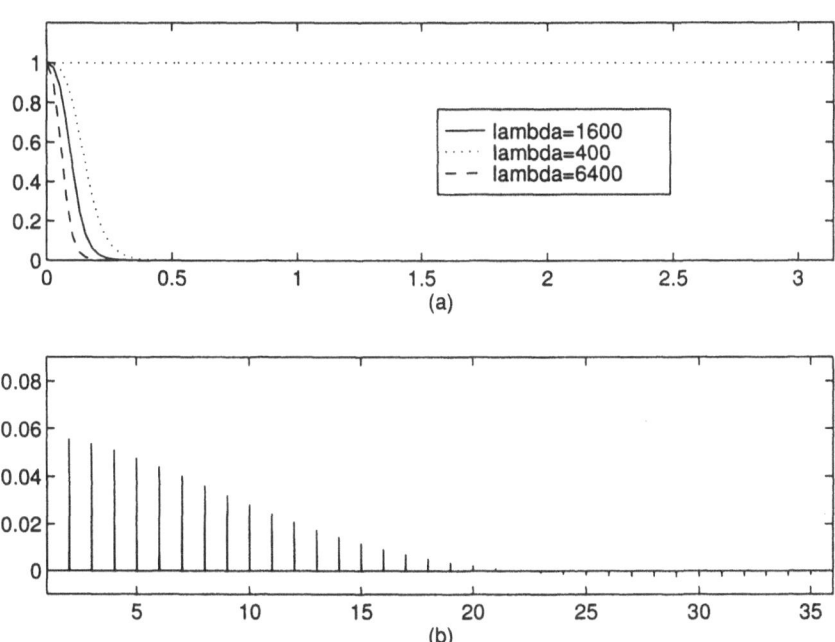

Figure 4.2. Hodrick–Prescott filter, trend: (a) squared gain–trend HP filter; (b) weights of the trend HP filter ($\lambda = 1600$).

For quarterly series, it has became overwhelming standard practice to use the value $\lambda = 1600$, originally proposed by Hodrick and Prescott (1980). The proposal was based on a somewhat mystifying belief that "a 5 percent cyclical component is moderately large, as is a one-eighth of 1 percent change in the growth rate in a quarter". (The arbitrariness of this assumption and the implicit risk of manipulating the length of the cycle has been pointed outoften.) For this standard value of λ, the squared gain was displayed in Figure 4.2(a). The figure shows that the filter will treat as trends movements with frequencies lower than, roughly, .17, and hence the trend filter will aim to capture cycles of periods longer than, roughly, nine years.

Figure 4.2(b) exhibits the weights of the (time-domain) filter for the case $\lambda = 1600$; since the filter is symmetric, only one side is displayed. One relevant feature, revealed by direct inspection, is that the filter extends, in a significant way, over a large number of periods. Therefore, estimators for recent quarters will remain preliminary for several years. This feature will imply unreliability of the trend estimator for the present and recent periods.

Up to now we have concentrated on the HP filter for estimating the trend, given by expression (4.2). Our main component of interest, however, is the cycle. If

- $\nu_{HP}^m(B, F) = $ HP filter that yields the trend estimator,

- $\nu_{HP}^c(B, F) = $ HP filter that yields the cycle estimator,

then, from the identity $x_t = \hat{m}_t + \hat{c}_t$, one obtains

$$\hat{c}_t = x_t - \hat{m}_t = [1 - \nu_{HP}^m(B, F)]x_t,$$

so that

$$\nu_{HP}^c(B, F) = 1 - \nu_{HP}^m(B, F). \tag{4.9}$$

Therefore, the weights of the HP filter to estimate the cycle are immediately obtained from the weights of the trend filter. If j denotes the exponent of B (or F, since the filters are symmetrical), then $\nu_{HP}^c(j) = \delta(j) - \nu_{HP}^m(j)$, where $\delta(j) = 1$ for $j = 0$ and 0 for $j \neq 0$. Similarly, in the frequency domain,

$$g_{\hat{c}}(\omega) = g_x(\omega) - g_{\hat{m}}(\omega) = [1 - G_{HP}^m(\omega)^2]g_x(\omega),$$

where $G_{HP}^m(\omega)$ is the gain of the HP trend filter. Therefore, the gain of the HP cycle filter is easily obtained from the trend one. Figure 4.3 displays the (time-domain) weights of the filter to estimate the cycle and its squared gain, for quarterly series and $\lambda = 1600$.

From Figure 4.3(a), it is seen that the HP filter for the cycle will pass all of the series variations associated with the quarterly seasonal frequencies $\omega = \pi/2$ and π. Given that seasonal variation should not contaminate the cycle, for seasonal series (as the vast majority of quarterly macroeconomic series, of relevance to business-cycle analysis, are) the HP filter has to be applied to seasonally adjusted series. In practice, the series has been typically seasonally adjusted with the program X11, or some of its variants (see Shiskin et al. (1967), Dagum (1980) or Findley et al. (1998)). Assuming no outliers, the complete X11 filter (applicable to historical estimators) can be seen as a linear, centered, and symmetric filter, designed to remove variation for the seasonal frequencies (see Figure 2.10(b)). Since all variants share the same basic default filter, for the rest of the book X11 denotes the default linear filter for an additive decomposition, as in Ghysels and Perron (1993). Furthermore, we always apply it, in the X11-ARIMA spirit, to the series extended at both ends with ARIMA backcasts and forecasts.

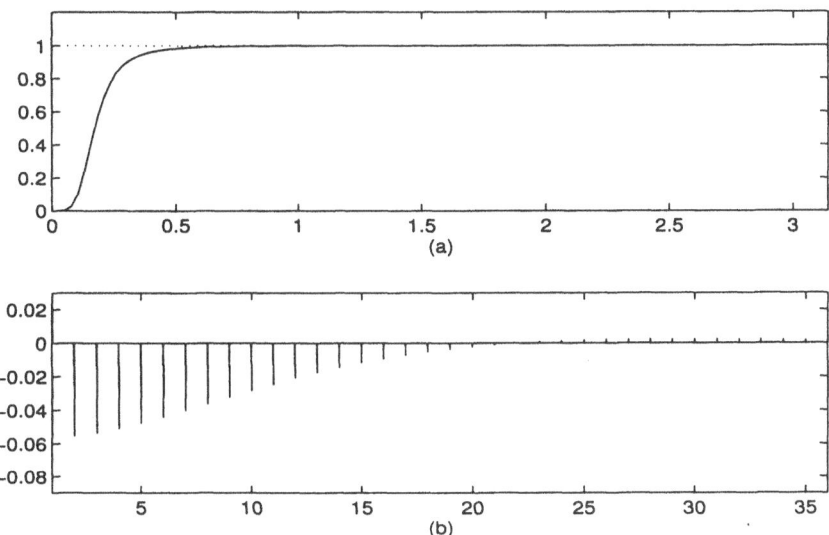

Figure 4.3. Hodrick–Prescott filter, cycle: (a) squared gain–cycle HP filter ($\lambda = 1600$); (b) weights of cycle HP filter ($\lambda = 1600$).

Letting x_t denote the observed (seasonal) series, it is first decomposed as in

$$x_t = n_t + s_t, \tag{4.10}$$

where n_t and s_t denote, respectively, the SA series and seasonal component. Estimating n_t with X11 yields

$$\hat{n}_t = \nu_{X11}(B, F)x_t, \tag{4.11}$$

where $\nu_{X11}(B, F)$ is the X11 filter. Then, applying the HP filter to $\hat{n}_t$, one obtains

$$\hat{c}_t = \nu^c_{HP}(B, F)\hat{n}_t = [\nu^c_{HP}(B, F)\nu_{X11}(B, F)]x_t,$$

and hence the full filter to estimate the cycle is the convolution of the X11 and the HP filters,

$$\nu^c_{HPX}(B, F) = \nu^c_{HP}\nu_{X11}(B, F). \tag{4.12}$$

(Because both the X11 and HP filters are symmetric, centered, and convergent, so will be their convolution.) For seasonal series, thus, the estimator of the cycle is obtained through

$$\hat{c}_t = \nu^c_{HPX}(B, F)x_t. \tag{4.13}$$

For $\lambda = 1600$, the weights of the ν^c_{HPX} filter, as well as its squared gain are displayed in Figure 4.4. Two features are worth mentioning:

(a) comparing Figures 4.4(b) and 4.3(b), it is seen that incorporation of the SA filter has a small effect on the (relevant) length of the filter.

(b) Inspection of part (a) of the figure shows that the combined X11-HP filter to estimate the cycle removes from the series cycles with periods larger than about 9 to 10 years, as well as cycles in a (fixed) neighborhood of the seasonal frequencies. In particular, high (nonseasonal) frequencies, with associated periods of less than a year, are passed to the cycle; these frequencies do not belong to the range of cyclical frequencies.

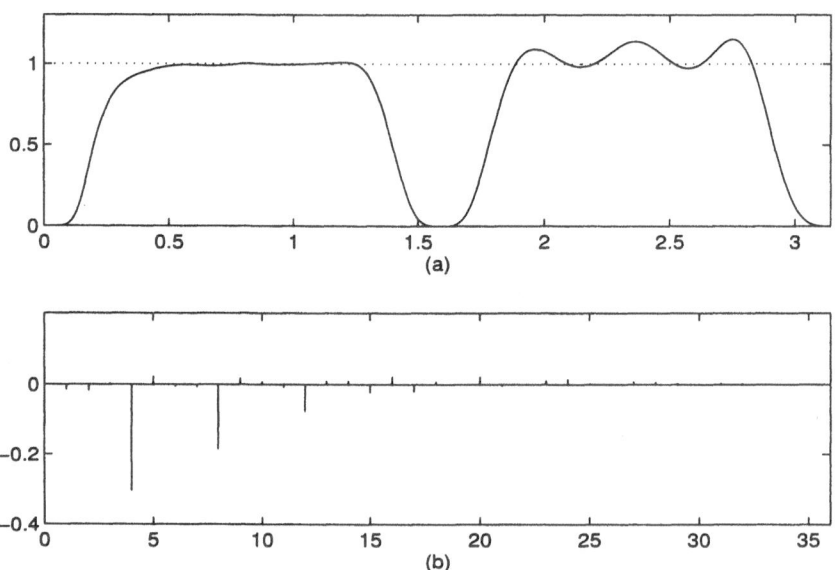

Figure 4.4. X11-HP cycle filter: (a) squared gain–cycle X11-HP filter ($\lambda = 1600$); (b) weights of cycle X11-HP filter ($\lambda = 1600$).

4.3 Some Criticisms and Discussion of the Hodrick–Prescott Filter

Looking at the basic characteristics of the HP filter in the previous section, we already detected several potential problems. First, the lack of a proper foundation for the derivation of λ can induce arbitrariness in the measurement of the cycle. (Still, the success of the $\lambda = 1600$ value may indicate that it roughly approximates the frequencies of interest to applied business-cycle analysts.) Second, we mentioned the slow convergence of the filter, which implies a long-lasting sequence of preliminary estimators, with the associated problem of revisions and unreliability of present and recent estimators (this would also affect detection of turning points). Third, we saw how the output of the filter could be contaminated with high frequency variation, which is contrary to the desirable properties of a cyclical filter. Fourth, when combined with X11 , the criticisms of X11 would also apply. An important element of this criticism is that the rigidity of the seasonal frequency holes in the squared gain of the filter can be too wide or too narrow for properly capturing the seasonality present in the actual series (see Maravall (1999)). We do not address this last criticism, but we do look in some detail at the others.

From a more general perspective, the HP (or X11-HP) filter is subject to the main criticism often made of ad hoc fixed filters, namely, that it may well produce spurious results (for the case of trends, see, e.g., Nelson and Kang (1981)). In particular, it is argued that the HP filter may induce cycles that are not present in the series. Harvey and Jaeger (1993) use a structural time series model with a cycle incorporated, and a trend modeled as

$$
\begin{aligned}
m_t &= m_{t-1} + \beta_{t-1} \\
\beta_t &= \beta_{t-1} + \epsilon_t, \qquad\qquad e_t \sim \text{niid}(0, V_\epsilon),
\end{aligned}
$$

which is easily seen to be identical to (4.7), to show how the cycle captured by their model can be unrelated to the HP cycle, which may seem an artificial imposition on the series. King and Rebelo (1993), Cogley and Nason (1995), and Maravall (1995) have further insisted on the spuriousness argument, and showed that application of the filter affects the series autocorrelation, distorting the relationship between them, and raising the possibility of spurious relationships.

The basic argument behind the spurious-effect criticism can be simply illustrated with a (quite general) example. Consider an unobserved component model given by a trend, a seasonal, and a cyclical component of the type

$$
\begin{aligned}
\nabla^d m_t &= \theta_m(B) a_{mt} \qquad (d > 0), & (4.14) \\
S s_t &= \theta_s(B) a_{st}, & (4.15)
\end{aligned}
$$

$$c_t = \theta_c(B)a_{ct}, \tag{4.16}$$

with the variables a_{mt}, a_{st}, and a_{ct} being mutually uncorrelated white noises with variances V_m, V_s, and V_c. The aggregate ARIMA model for the series

$$x_t = m_t + s_t + c_t \tag{4.17}$$

is then of the type

$$\nabla^{d-1}\,\nabla_4 x_t = \theta(B)a_t,$$

where $\theta(B)a_t$ satisfies the constraint (similar to (3.47))

$$\theta(B)a_t = S\theta_m(B)a_{mt} + \nabla^d\theta_s(B)a_{st} + \nabla^d S\theta_c(B)a_{ct}.$$

The model consisting of equations (4.14) to (4.17) includes many particular cases, such as the model-based approximations to X11 of Cleveland and Tiao (1976) and Burridge and Wallis (1984), the model-based representation of the HP filter given by King and Rebelo (1993), and many structural time series models actually proposed (see, e.g., Harvey and Todd (1983), Engle (1978) and Gersh and Kitagawa (1983)). The MMSE of the cycle is given by (see Section 3.4)

$$\hat{c}_t = \left[k_c \frac{\theta_c(B)\nabla^d S\,\theta_c(F)\bar{\nabla}^d\bar{S}}{\theta(B)\qquad\theta(F)} \right] x_t, \tag{4.18}$$

where $k_c = V_c/V_a$, and $\bar{\nabla} = 1 - F, \bar{S} = 1 + F + F^2 + F^3$. Replacing x_t by its ARIMA model, after simplifications, $\hat{c}_t$ can be expressed as

$$\hat{c}_t = k_c \frac{\theta_c(B)\theta_c(F)}{\theta(F)} U(F)a_t, \tag{4.19}$$

where

$$U(F) = \bar{\nabla}^d\bar{S}.$$

Considering that $U(F)$ contains unit roots for $\omega = 0, \omega = \pi/2$, and $\omega = \pi$, the spectrum of (4.19) will satisfy

$$g_{\hat{c}}(0) = g_{\hat{c}}(\pi/2) = g_{\hat{c}}(\pi) = 0$$

and, assuming $\theta_c(B)$ has no unit roots for nonzero or nonseasonal frequencies, $g_{\hat{c}}(\omega) > 0$ for other values of ω. Therefore the shape of $g_{\hat{c}}(\omega)$ has to always consist of two "hills", one between $\omega = 0$ and $\pi/2$, and the other between $\omega = \pi/2$ and π.

Let us take now the filter in brackets in (4.18) as a fixed filter (many fixed filters for detrending and SA fit into this model representation, at least as a close approximation; see Gómez and Maravall (2000b)). Assume the series x_t is white noise (hence no cycle is present in the series). The spectrum of the filtered series displays a peak in the range of cyclical frequencies (the one that corresponds to the first hill, between $\omega = 0$ and $\pi/2$). This peak could be interpreted as evidence of the existence of a cycle; yet, by construction, the series contained no cycle. A particular case of the previous general example is of help. Consider the observed quarterly series x_t, which is the sum of three uncorrelated components: a trend (m_t), a seasonal component (s_t), and a white noise irregular component (c_t). The models for m_t and s_t are given by

$$\nabla^2 m_t = a_{mt},$$
$$S s_t = a_{st}$$

(thus the trend has the same model as the HP trend, and the seasonal component is as in the basic STS model of Harvey and Todd (1983)). Furthermore, assume the following values for the component innovations: $V_m = V_c/1600$, $V_c = 1$.

By construction, the SA and detrended series is c_t, hence white noise. From (4.19), the spectrum of the filtered series $\hat{c}_t$ can be computed; it is shown in Figure 4.5. The spectrum of $\hat{c}_t$ displays an important peak for the five-year cyclical frequency. If an AR model is fit to series generated from this model, this cyclical peak is likely to be captured. The captured cycle would be entirely induced by the filtering, since by construction the white noise input series were not cyclical. In this sense, the cycle detected in the filtered series would be spurious.

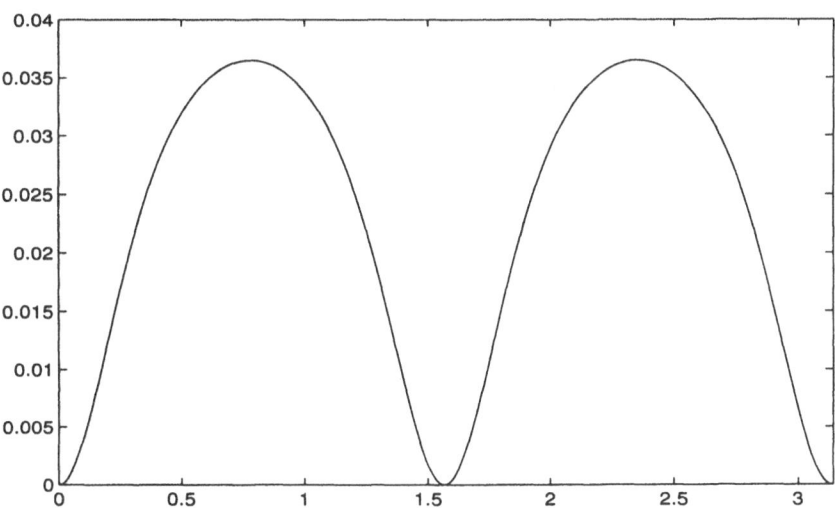

Figure 4.5. spectrum of the white noise component estimator.

Of course, a sensible time series analyst would not apply a detrending and SA filter to white noise. Thus the importance of spuriousness in practice may be moderate. A reflection, however, that is relevant to the spuriousness issue is the following. After the influential work of Box and Jenkins (1970), time series analysts had become used to routinely remove trends by differencing the series. This was the first step in the construction of parsimonious ARIMA models; and for these models, the effect of differencing often extends over the range of cyclical frequencies, and beyond. Moreover, the effect of differencing on seasonal frequencies is possibly related to the difficulty in identifying cyclical AR structures in the models identified through the Box–Jenkins standard procedure (with some notable exceptions, such as Jenkins (1975) or Crafts et al. (1989)). The fact that differencing tends to remove more variation from the series than the HP filter, and that part of this variation could be of cyclical interest was pointed out by Singleton (1988) and is also discussed in Canova (1998). In terms of the analysis of Section 3.6, ARIMA models tend to capture short-term trends (or trend-cycle components), while the HP trend is designed as a longer-term trend. We come back to the spuriousness criticism later, and our final verdict is that the criticism is, often, unwarranted.

A limitation of the HP filter, often pointed out, is the poor behavior of the cyclical estimator at the end of the series. As was mentioned earlier, the length of the filter implies a long revision period, and hence is likely to imply instability in recent estimators. To avoid this problem, Baxter and King (1999) have suggested a filter, similar to the HP one, but with better time aggregation properties (an issue we do not address in this work) that does not provide estimators of the cycle for the end periods. Given that we consider endpoint estimation a crucial issue, we stick to the HP filter. As Canova (1998) and Cogley and Nason (1995) have shown, the "stylized facts" of the estimated cycle are strongly affected by the method used for detrending; thus a positive feature of the generalized use of the HP filter is that it has brought homogeneity in method, so that the effect of the filter has been stabilized.

In the next two chapters we proceed to discuss systematically and in more detail the major limitations of the HP filter. To do that, it proves helpful to use an alternative derivation of the filter.

4.4   The Hodrick–Prescott Filter as a Wiener–Kolmogorov Filter

4.4.1   An Alternative Representation

We provide an alternative representation of the HP filter, based on the WK one, that provides an efficient and simple computational algorithm and, in addition, turns out to be very useful for analytical discussion. Starting with the King and Rebelo model-based interpretation,

$$x_t = m_t + c_t,$$
$$\nabla^2 m_t = a_{mt},$$

with c_t and a_{mt} uncorrelated white noises, by noticing that $\nabla^2 x_t = a_{mt} + \nabla^2 c_t$, it follows that x_t will follow an IMA(2,2) model of the type

$$\nabla^2 x_t = (1 + \theta_1 B + \theta_2 B^2) b_t = \theta_{HP}(B) b_t, \qquad (4.20)$$

where b_t are the innovations in the x_t series. The variance of b_t, V_b, and the θ_1, θ_2-parameters are found by factorizing the spectrum from the identity

$$(1 + \theta_1 B + \theta_2 B^2) b_t = a_{mt} + \nabla^2 c_t, \qquad (4.21)$$

or, as seen in the next section, by solving the covariance equations associated with (4.21). As an example, for quarterly series and the standard value $\lambda = 1600$ it is found that

$$\theta_{HP}(B) = 1 - 1.77709 B + .79944 B^2; \qquad V_b = 2001.4. \qquad (4.22)$$

For an infinite realization of the series, the MMSE estimator of m_t is given by (3.27) which, for the present model, simplifies into

$$\hat{m}_t = k_{m(HP)} \frac{1}{\theta_{HP}(B)\theta_{HP}(F)} x_t = \nu^m_{HP}(B, F) x_t, \qquad (4.23)$$

where $k_{m(HP)} = V_m/V_b$. The filter $\nu^m_{HP}(B, F)$ is symmetrical and, since (4.21) implies that $\theta_{HP}(B)$ is invertible, also convergent. As seen in Section 3.4, for a finite series, expression (4.23) can still be applied, with x_t replaced by the series extended with forecasts and backcasts. A simple and efficient algorithm to apply the filter is detailed in the next section. For the estimator of the cycle, again applying (3.27) to our case,

$$\hat{c}_t = \nu^c_{HP}(B, F) x_t = \left[k_{c(HP)} \frac{\nabla^2 \bar{\nabla}^2}{\theta_{HP}(B)\theta_{HP}(F)} \right] x_t, \qquad (4.24)$$

where $k_{c(HP)} = V_c/V_b$, and, as before, a bar over an operator denotes the same operator with B replaced by F. It is straightforward to verify that the two filters $\nu^m_{HP}(B, F)$ and $\nu^c_{HP}(B, F)$ satisfy equation (4.9). When properly applied, the Danthine and Girardin, the Kalman filter, and the WK solutions are numerically identical (see Gómez (1999)). The last two are considerably more efficient than the first, and can be applied to series of any length. The WK filter turns out to be convenient for analytical discussion.

It proves useful to present the frequency domain version of the filter (4.24) (i.e., the gain of the filter), denoted $\tilde{\nu}^c_{HP}(\omega)$. Using (4.21), the filter (4.24) can be rewritten as

$$\nu^c_{HP}(B, F) = \frac{\nabla^2 \bar{\nabla}^2 V_c}{V_m + \nabla^2 \bar{\nabla}^2 V_c} = \frac{\nabla^2 \bar{\nabla}^2}{\lambda^{-1} + \nabla^2 \bar{\nabla}^2}$$

(where $\lambda = V_c/V_m$) with Fourier transform given by (see Section 2.5),

$$\tilde{\nu}_{HP}^c(\omega) = \frac{4(1 - \cos\omega)^2}{\lambda^{-1} + 4(1 - \cos\omega)^2} \quad . \tag{4.25}$$

4.4.2 Derivation of the Filter

(a) The model for the aggregate series.

In order to apply the filter (4.24), we need to know the polynomial $\theta_{HP}(B)$. We concluded that the model for x_t is an IMA(2,2) model, which we represent by (4.20). We need to compute θ_1, θ_2, and the variance of the innovation b_t for a given value of the HP filter parameter λ. Equating the AGF of the two sides of equation (4.21), a system of three nonlinear equations is obtained, one for the variance, and one for each of the lag-1 and lag-2 autocovariances. From Appendix A in Maravall and Mathis (1994), the system is trivially solved as follows (all square roots in the expressions are taken with their positive sign).

Compute sequentially:

(1) $a = 2$, $b = 1/\sqrt{\lambda}$, $k = -b^2$, $s = 2ab$;

(2) $z = \sqrt{\frac{1}{2\lambda}(1 + \sqrt{1 + 16\lambda})}$; $r = s/2z$;

(3) $m_1 = (-a + r)/2$, $n_1 = (z - b)/2$;

(4) $m_2 = (-a - r)/2$, $n_2 = (-z - b)/2$;

(5) of the two numbers $(m_1 + in_1)$ and $(m_2 + in_2)$ pick up the one with smallest modulus. Let this number be

$$R = M + iN;$$

(6) $\theta_1 = 2M$, $\theta_2 = M^2 + N^2$, $V_b = (1 + 6\lambda)/(1 + \theta_1^2 + \theta_2^2)$.

An alternative, equivalent, procedure to obtain θ_1, θ_2, and V_b is the following.

- Compute the variance and covariances of the term in the right-hand side of (4.21),

$$\begin{aligned}
\gamma_0 &= V_m + 6V_c; \\
\gamma_1 &= -4V_c; \\
\gamma_2 &= V_c.
\end{aligned}$$

- Find the roots of the polynomial

$$p(x) = \gamma_2 + \gamma_1 x + \gamma_0 x^2 + \gamma_1 x^3 + \gamma_2 x^4.$$

- Choose the pair of complex conjugate roots with modulus < 1. Let this pair be

$$R = M \pm Ni.$$

- Then, $\theta_1 = 2M, \theta_2 = M^2 + N^2, V_b = (1 + 6\lambda)/(1 + \theta_1^2 + \theta_2^2)$.

In this way, given the value of λ, the HP filter parameter, the model for the aggregate series x_t is fully specified.

(b) Application of the Burman–Wilson algorithm.

As mentioned in Section 4.1, without loss of generality, we can set $V_m = 1, V_c = \lambda$; thus

$$k_{m(HP)} = V_m/V_b = 1/V_b; \qquad k_{c(HP)} = V_c/V_b = \lambda/V_b.$$

To simplify notation, we remove the subindex HP from the polynomials in B and the k-ratios; thus, for the rest of the chapter, $\theta_{HP}(B)$ becomes $\theta(B)$, $k_{m(HP)}$ becomes k_m, and so on.

We can now apply an adaptation of the Burman–Wilson algorithm (Burman, 1980) to compute the HP trend with the Wiener–Kolmogorov filter. Due to its symmetry, the WK filter to estimate m_t, given by (4.23), can be expressed as

$$\nu(B, F) = \frac{k_m}{\theta(B)\theta(F)} = k_m \left[\frac{G(B)}{\theta(B)} + \frac{G(F)}{\theta(F)} \right], \qquad (4.26)$$

where $G(B) = g_0 + g_1 B + g_2 B^2$. Removing denominators in the above identity and equating the coefficients of the terms in B^0, B, and B^2, yields a system of equations that can be solved for g_0, g_1, and g_2. Specifically, if

$$\mathbf{A} = \begin{bmatrix} 1 & 0 & 0 \\ \theta_1 & 1 & 0 \\ \theta_2 & \theta_1 & 1 \end{bmatrix} + \begin{bmatrix} 0 & 0 & \theta_2 \\ 0 & \theta_2 & \theta_1 \\ \theta_2 & \theta_1 & 1 \end{bmatrix},$$

the solution is given by

$$[g_2 \quad g_1 \quad g_0]' = A^{-1}[0 \quad 0 \quad 1]'. \qquad (4.27)$$

Using (4.26), write $\hat{m}_t = \nu(B, F)x_t$ as

$$\hat{m}_t = k_m[x_t^B + x_t^F], \qquad (4.28)$$

where

$$x_t^B = [G(B)/\theta(B)]x_t, \qquad (4.29)$$

$$x_t^F = [G(F)/\theta(F)]x_t. \qquad (4.30)$$

We need four backcasts and four forecasts of x_t; they can be computed in the usual Box–Jenkins way through model (4.20) for the aggregate series.

(i) Computation of x_t^F. Differencing (4.30) twice, and considering (4.20), yields

$$
\nabla^2 x_t^F = \frac{G(F)}{\theta(F)}\nabla^2 x_t = \frac{G(F)}{\theta(F)}\theta(B)a_t
$$
$$
= (\psi_0 + \psi_1 F + \psi_2 F^2 + \ldots)(1 + \theta_1 B + \theta_2 B^2)a_t,
$$

or

$$
x_t^F - 2x_{t-1}^F + x_{t-2}^F = (\alpha_2 B^2 + \alpha_1 B + \alpha_0 + \sum_{j=1}^{\infty}\alpha_{-j}F^j)a_t. \qquad (4.31)
$$

Taking expectations at time T in both sides of (4.31), and assuming (4.20) is the correct model, $E_T[a_{T+k}] = 0$ for $k > 0$, so that for $t = T + 3$ and $T + 4$, it is obtained that

$$
x_{T+3}^F - 2x_{T+2}^F + x_{T+1}^F = 0; \qquad (4.32)
$$
$$
x_{T+4}^F - 2x_{T+3}^F + x_{T+2}^F = 0. \qquad (4.33)
$$

Let x_t include the four forecasts of the series, and compute the auxiliary series $y_t = G(F)x_t$, $t = 1, \ldots, T + 2$. From (4.30), $\theta(F)x_t^F = y_t$, or, for $t = T + 1, T + 2$,

$$
x_{T+1}^F + \theta_1 x_{T+2}^F + \theta_2 x_{T+3}^F = y_{T+1}; \qquad (4.34)
$$
$$
x_{T+2}^F + \theta_1 x_{T+3}^F + \theta_2 x_{T+4}^F = y_{T+2}. \qquad (4.35)
$$

The system of four equations (4.32) to (4.35) can be solved for $x_{T+1}^F, \ldots, x_{T+4}^F$. The remaining x_T^F are computed recursively through

$$
x_t^F = -\theta_1 x_{t+1}^F - \theta_2 x_{t+2}^F + y_t; \qquad t = T, \ldots, 1.
$$

(ii) Computation of x_t^B. Proceeding in a symmetric manner, compute the auxiliary series $z_t = G(B)x_t$, where x_t now includes four backcasts at the beginning and four forecasts at the end. From (4.29),

$$
(1 - F)^2 x_t^B = \frac{G(B)}{\theta(B)}\theta(F)e_t, \qquad (4.36)
$$

where $(1 - F)^2 x_t = \theta(F)e_t$, so that e_t is the forward residual. These residuals will now satisfy $E_T[e_{T-k}] = 0$ for $k > 0$. Proceeding as before, taking conditional expectations in (4.36) for $t = -3$ and -2 yields

$$
x_{-3}^B - 2x_{-2}^B + x_{-1}^B = 0, \qquad (4.37)
$$
$$
x_{-2}^B - 2x_{-1}^B + x_0^B = 0, \qquad (4.38)
$$

and, from (4.29), $\theta(B)x_t^B = z_t$ for $t = -1, 0, 1, \ldots, T + 4$. Therefore

$$
x_{-1}^B + \theta_1 x_{-2}^B + \theta_2 x_{-3}^B = z_{-1}, \qquad (4.39)
$$
$$
x_0^B + \theta_1 x_{-1}^B + \theta_2 x_{-2}^B = z_0. \qquad (4.40)
$$

The system consisting of the four equations (4.37) to (4.40) can now be solved for $x_{-3}^B, x_{-2}^B, x_{-1}^B, x_0^B$. The rest of the x_t^B are obtained recursively from

$$x_t^B = -\theta_1 x_{t-1}^B - \theta_2 x_{t-2}^B + z_t; \qquad t = 1, \ldots, T.$$

Finally having obtained x_t^B and x_t^F, the estimator $\hat{m}_t$ is obtained through (4.28). Notice that the procedure automatically provides four forecasts.

4.4.3 The Algorithm

We consider the standard quarterly case with $\lambda = 1600$, for which $\theta_{HP}(B)$ and V_b are given by (4.22). From (4.27) it is found that $g_0 = -44.954$, $g_1 = 11.141$, and $g_2 = 56.235$. The matrix of coefficients in the two sets of equations (4.32) to (4.35) and (4.37) to (4.40) is

$$\begin{bmatrix} 1 & -2 & 1 & 0 \\ 0 & 1 & -2 & 1 \\ 1 & -1.7771 & .7994 & 0 \\ 0 & 1 & -1.7771 & .7994 \end{bmatrix}.$$

Denote by **H** the inverse of this matrix. Let $x_t = [x_1, \ldots, x_T]$ be the series for which we wish to estimate the HP trend m_t, and extend the series at both ends with four forecasts and four backcasts, computed with model (4.20). Then the algorithm that yields $\hat{m}_t$ is the following.

Step I. For $t = 1, \ldots, T + 2$, compute (using four forecasts)

$$y_t = g_0 x_t + g_1 x_{t+1} + g_2 x_{t+2},$$

$$[x_{T+1}^F, \ldots, x_{T+4}^F]' = H[0, 0, y_{T+1}, y_{T+2}]',$$

and, for $t = T, \ldots, 1$, obtain recursively

$$x_t^F = -\theta_1 x_{t+1}^F - \theta_2 x_{t+2}^F + y_t.$$

Step II. For $t = -1, 0, 1, \ldots, T + 4$ compute (using four backcasts)

$$z_t = g_0 x_t + g_1 x_{t-1} + g_2 x_{t-2},$$

$$[x_0^B, x_{-1}^B, x_{-2}^B, x_{-3}^B]' = H[0, 0, z_0, z_{-1}]',$$

and, for $t = 1, \ldots, T + 4$, obtain recursively

$$x_t^B = -\theta_1 x_{t-1}^B - \theta_2 x_{t-2}^B + z_t.$$

Step III. For $t = 1, \ldots, T + 4$, obtain

$$\hat{m}_{t|T} = k_m [x_t^F + x_t^B].$$

This yields the trend estimated for the sample period $t = 1, \ldots, T$, and forecasted for the periods $t = T + 1, \ldots, T + 4$. The algorithm consists of a few convolutions and some minor matrix multiplications. It is fast and reliable, even for a series with (say) a million observations.

4.4.4   A Note on Computation

The previous derivation corresponds to the usual HP filter, and clearly indicates a vulnerable point that accounts, at least partly, for the poor behavior at the end of the series. The algorithm needs four forecasts and four backcasts of the series x_t. The standard HP filter implies that these extensions are computed using model (4.20) but, in general, the observed series will not follow (4.20) but some other ARIMA model, which we assume has been identified for the series. In that case, it is only for the later model that the conditions $E_T(a_{T+k}) = 0$ and $E_T(e_{T-k}) = 0$ for $k > 0$, necessary for the validity of the two systems of equations (4.32) to (4.35) and (4.37) to (4.40), hold; for the "residuals" produced by model (4.20) these conditions will not be satisfied. Therefore, an obvious improvement in the filter will be achieved simply if the forecasts and backcasts used are those obtained with the correctly specified model. We come back to this issue later.

The procedure assumes that the models are linear stochastic processes, in which case the optimal estimators are obtained with the linear filters we derived earlier. As mentioned in Section 3.4, in practice, many series may need prior treatment before the linearity assumption can be made. Program TRAMO (time series regression with ARIMA noise, missing values, and outliers) can be used for automatic (or manual) pretreatment. The program outputs the series for which an ARIMA model can be assumed. This ARIMA model can then be decomposed (automatically or manually) with the SEATS program and the SA series, and the trend-cycle estimator, together with their forecasts and backcasts can be obtained. Running SEATS again on the extended SA or trend-cycle series, with the fixed specifications of the HP filter (for the standard quarterly series given by (4.20) and (4.22)), the estimator of the cycle, as well as its forecasts, can be obtained (this procedure of business cycle estimation will be included in the next version of SEATS).

Both (documented) programs can be freely downloaded from the site http://www. bde.es, and are described in Gómez and Maravall (1996). They can also be supplied upon request.

5.
Some Basic Limitations of the Hodrick–Prescott Filter

5.1 Endpoint Estimation and Revisions

5.1.1 Preliminary Estimation and Revisions

In Section 4.1 we presented the HP filter as a symmetric two-sided filter. Given that the concurrent estimator is a projection on a subset of the set of information that provides the final estimator, the latter cannot be less efficient. Besides, concurrent estimators, obtained with a one-sided filter, induce phase effects that distort the timing of events and harm early detection of turning points.

Although the discussion applies equally to any preliminary estimator, we center attention on concurrent estimation, that is, on the estimator of the cycle for the last observed period to be denoted $\hat{c}_{T|T}$. As new periods are observed the estimator is revised to $\hat{c}_{T|T+1}, \hat{c}_{T|T+2}, \ldots$ until it converges to the final estimator $\hat{c}_T$ (see Section 3.4). The difference between the final and concurrent estimators measures the total revision the concurrent estimator will undergo, and can be interpreted as a measurement error contained in the concurrent estimator.

Although the poor behavior of the HP filter for recent periods has often been pointed out (see Baxter and King (1999)), the revisions implied by HP filtering have not been analyzed. Two main features of the revision are of interest: its magnitude, and the duration of the revision process (i.e., the value of L in (3.33) at which the filter can be safely truncated and $\hat{c}_{T|T+k}$ has, in practice, converged). To look at these features we use the

WK version of the HP filter for the cycle, given by (4.24), and proceed as follows.

Assume the observed series follows the ARIMA model

$$\phi(B)\nabla^d x_t = \theta(B)a_t, \tag{5.1}$$

with $0 \le d \le 4$, $\phi(B)$ stationary, and $\theta(B)$ invertible. Combining (4.24) with (5.1), it is possible to express the estimator of the cycle as a linear filter in the innovations of the observed series a_t, and obtain an expression of the type (3.38). After simplification, it is found that

$$\hat{c}_t = \xi(B, F)a_t = \left[k_{c(HP)} \frac{\nabla^{4-d}\theta(B)}{\theta_{HP}(B)\phi(B)\theta_{HP}(F)} \right] a_{t+2}. \tag{5.2}$$

Given that $d \le 4$, $\theta_{HP}(B)$ is invertible, and $\phi(B)$ stationary, the filter $\xi(B, F)$ converges in F and in B, and $\hat{c}_t$ is a stationary process. From expression (5.2) the spectrum of $\hat{c}_t$ can be easily obtained. Because it is empirically very unlikely that $d \ge 4$, the factor ∇^{4-d} will induce a zero in that spectrum for the zero frequency, and $\hat{c}_t$ will be a noninvertible series.

Direct inspection of expression (5.2) shows that the spectrum of $\hat{c}_t$ is determined, in part, by the structure of the HP filter and, in part, by the dynamic structure of the observed series. Expression (5.2) can be rewritten as

$$\hat{c}_t = \xi^-(B)a_t + \xi^+(F)a_{t+1}, \tag{5.3}$$

where

$$\xi^-(B) = \sum_{j \ge 0} \xi_{-j} B^j,$$

and

$$\xi^+(F) = \sum_{j \ge 0} \xi_j F^j$$

are convergent polynomials. The first one contains the effect of the innovations up to and including period t, and the second one includes the effect of innovations posterior to period t. Because

$$E_t(a_{t-j}) = a_{t-j} \qquad \text{when } j \ge 0,$$

$$E_t(a_{t-j}) = 0 \qquad \text{when } j < 0,$$

the concurrent estimator, equal to the expectation at time t of c_t, is given by the first term in the right-hand side of the equation. The revision in the concurrent estimator is thus given by

$$r_{t|t} = \xi^+(F)a_{t+1} = \sum_{j=1}^{h} \xi_j a_{t+j}, \tag{5.4}$$

where the second equality uses a finite approximation based on the convergence of $\xi^+(F)$. From (5.4), it is straightforward to compute the variance and autocorrelations of the revision process. (We have focused on the concurrent estimator; the analysis is trivially extended to any preliminary estimator $\hat{c}_{t|T}$.)

Although the filter $\nu_{HP}^c(B, F)$ is fixed, the coefficients of the forward filter $\xi^+(F)$ also depend on the ARIMA model for the observed series. Without loss of generality, we set $Var(a_t) = 1$, so that the variance of the revision

$$Var(r_{t|t}) = \sum_{j=1}^{h}(\xi_j)^2 \qquad (5.5)$$

is then expressed as a fraction of the variance of the series innovation V_a. Table 5.1 exhibits, for three models, the size of the revision and the number of periods needed for the concurrent estimator to converge to the final one (convergence is defined in practice as having removed more than 95% of the revision variance). The first example is for the case of a white noise series ($x_t = a_t$), and thus illustrates the "pure filter" effect. The second is the random walk model $\nabla x_t = a_t$, and the third example is the model for which the HP filter is optimal, namely, (4.20) and (4.22). The three examples represent, thus, an I(0), I(1), and I(2) variable, respectively.

Table 5.1. Revisions Implied by the HP Filter

Variable	Standard Deviation of Revision in Concurrent Estimator (as a Percent of σ_a)	Periods Needed for Convergence
White noise	13.9	12
Random walk	91.3	9
HP-IMA(2,2)	34.0	9

Even for the case in which the model is the one associated with optimality of the filter, the size of the revision is not negligible (approximately 34% of the one-period-ahead forecast error) and the revision period lasts in practice more than two years in all cases.

As already mentioned, the HP filter is often applied to X11-SA series, and the convolution of the two filters was earlier denoted $\nu_{HPX}(B, F)$ in expression (4.12). It is well known that X11, another two-sided filter, also produces revisions. Therefore, the revisions associated with the filter ν_{HPX} reflect the combined effect of the two filters. To look at this effect, we consider the case of white noise input. Proceeding as before, it is found that 95% of the revision variance disappears after 13 quarters, and that the revision standard deviation is 1/4 of the standard deviation of a_t. Comparing these results with the first row of Table 5.1, the addition of X11 increases the revision size, but the revision period barely changes.

To illustrate the revisions for series of more applied relevance we select the so-called airline model, discussed in Section 3.2, given by expression (3.19). The model fits well many series with trend and seasonality, and has became a standard example. For the most relevant range for the parameters θ_1 and θ_4, Table 5.2 presents the fraction $\sigma(\text{revision})/\sigma(a_t)$ and the number of periods (τ) needed for a 95% convergence in variance. The standard deviation of the revision represents between .4 and 1.5 of $\sigma(a_t)$, and convergence takes, roughly, between two and five years. Given that θ_1 close to -1 implies very stable trends, while θ_4 close to -1 implies very stable seasonals, what Table 5.2 shows is that series with highly moving trends and seasonals are subject to bigger, longer-lasting, revisions. It is worth pointing out that, for the range of values most often found in practice (see the study on more than 14,000 real series from 17 countries in Fischer and Planas (1999)) which is the bottom right corner, the revision period is equal to 9 quarters.

Table 5.2. Revisions Implied by the HP-X11 Filter.

	$\theta_4 = 0$		$\theta_4 = -.2$		$\theta_4 = -.4$		$\theta_4 = -.6$		$\theta_4 = -.8$	
	σ_r/σ_a	τ	σ_r/σ_a	τ	σ_r/σ_a	τ	σ_r/σ_a	τ	σ_r/σ_a	τ
$\theta_1 = .4$	1.53	19	1.44	18	1.36	17	1.28	9	1.21	9
$\theta_1 = .2$	1.34	19	1.26	18	1.18	17	1.12	9	1.06	9
$\theta_1 = 0$	1.15	19	1.08	18	1.02	16	0.96	9	0.90	9
$\theta_1 = -.2$	0.97	19	0.91	18	0.85	15	0.80	9	0.76	9
$\theta_1 = -.4$	0.79	18	0.74	17	0.70	14	0.65	9	0.61	9
$\theta_1 = -.6$	0.64	15	0.60	14	0.55	9	0.51	9	0.47	9
$\theta_1 = -.8$	0.52	9	0.48	9	0.44	9	0.40	9	0.36	9

5.1.2 An Example

An application, that is also used in later sections, completes the discussion. We consider four quarterly Spanish short-term economic indicators that can be reasonably suspected of being related to the business cycle. The series are the industrial production index (IPI), cement consumption (CC), car registration (CR), and airline passengers (AP) for the period 1972/1 to 1997/4, and contain 104 observations. (For the IPI series, the first 12 observations were missing and the period was completed using backcasts.) The series were log transformed (following proper comparison of the BIC criteria), and the application is discussed for the additive decomposition of the logs. Moreover, so as to facilitate comparisons, we standardize the four logged series to have zero mean and unit variance. The four series are represented in Figure 5.1; their trend and seasonal behavior are clearly discernible.

ARIMA modeling of the four series produced similar results: the models were of the type (3.19) (i.e., of the airline type) and a summary of results is given in Table 5.3; none of the series appeared to be in need of outlier adjustment. (Estimation was made with the TRAMO program run in an automatic mode.) Using the ARIMA models to extend the series, the HP ($\lambda = 1600$) filter was applied to the X11-SA series, and the four trends and four cycles obtained are displayed in Figures 5.2 and 5.3. For the series CC the short-term contribution of the cyclical variation is largest; it is smallest for the series AP.

Table 5.3. Summary of ARIMA Estimation Results

	Parameter Estimates		Residual	BL Test	Normality
	θ_1	θ_4	Variance V_a	$Q(< \chi^2_{14})$	$N(< \chi^2_2)$
CC	-.405	-.957	.175	18.4	.32
IPI	-.299	-.721	.054	23.3	.14
CR	-.387	-.760	.156	18.7	.79
AP	-.392	-.762	.017	21.1	2.76

Short-term monitoring focuses on recent periods, that is, on the concurrent estimator and its first revisions and, in fact, it is often the case that the HP filter is treated as a one-sided filter (see Prescott (1986)). We have argued before that the two-sided final estimator is preferable. The example illustrates why, even for the penalty-function derivation, the superiority of the final estimator is clear. Using the first and last 22 periods for safe convergence of the X11 and the HP filters, we obtained the sequence of concurrent and final estimators of the trend and cycle for the 60 central periods of the four series. Then, we evaluated the standard loss function of the HP filter, given by (4.5), for the concurrent and final estimators of the trend and cycle; the results are given in Table 5.4.

Table 5.4. HP Loss-Function for Concurrent and Final Estimator

	Concurrent Estimator	Final Estimator
CC	624.7	13.3
IPI	172.8	2.9
CR	513.4	11.9
AP	43.2	1.0

The improvement achieved by using final estimators instead of concurrent ones is indeed large. Figures 5.4 and 5.5 compare the series of concurrent and final estimators, for the trend and cycle, respectively. The differences are remarkable, and a clear phase effect in the concurrent estimator can be observed for the four series. Figure 5.6 illustrates the evolution of the cycle estimator from concurrent to final for three periods ($t = 61, 65$, and 70,) and Table 5.5 compares the numerical values of the two estimators for the three periods. Considering that the original series were standardized ($\mu = 0, \sigma = 1$), the revision in the estimator of the cycle is, in general, important. For the four series, 95% of the revision is completed in nine quarters, in agreement with the results of Table 5.2. The standard error of the revision is on the order of $.5\sigma_a$, certainly nonnegligible.

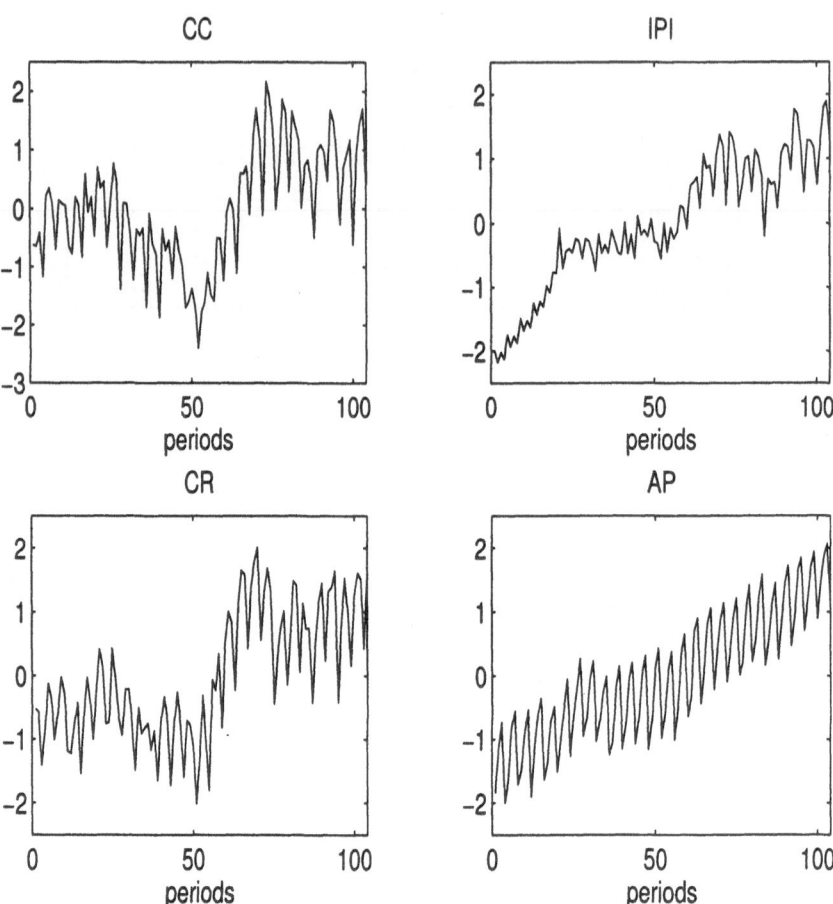

Figure 5.1. Short-term economic indicators: original series.

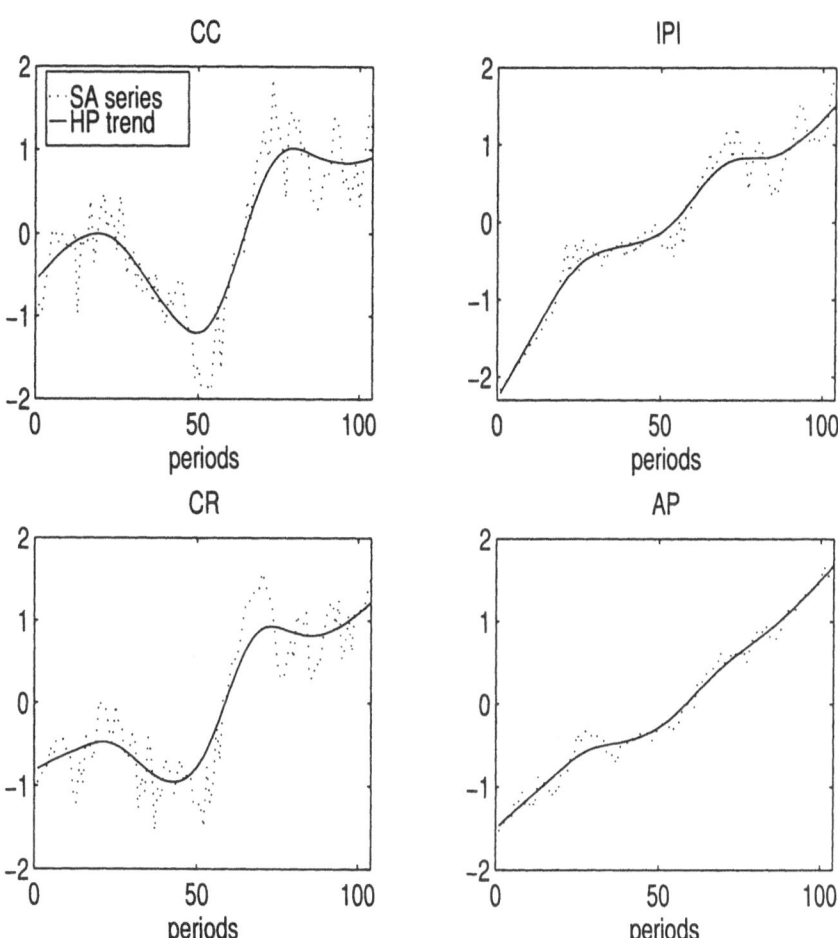

Figure 5.2. X11-SA series and HP trend.

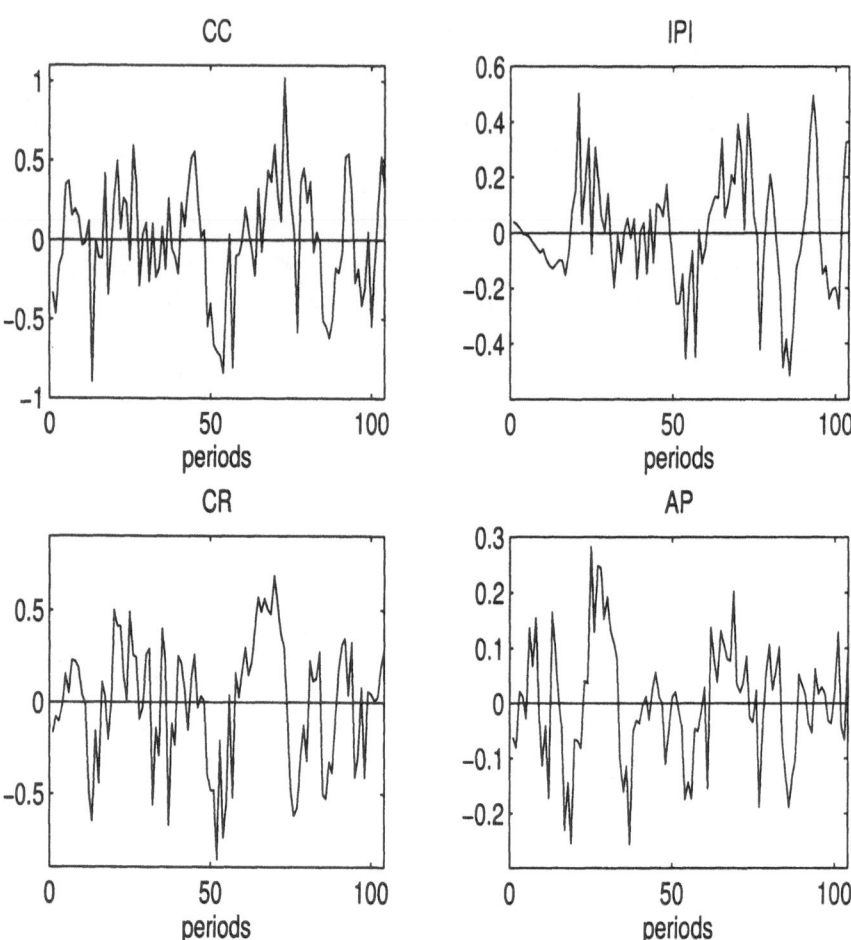

Figure 5.3. X11-HP cycles.

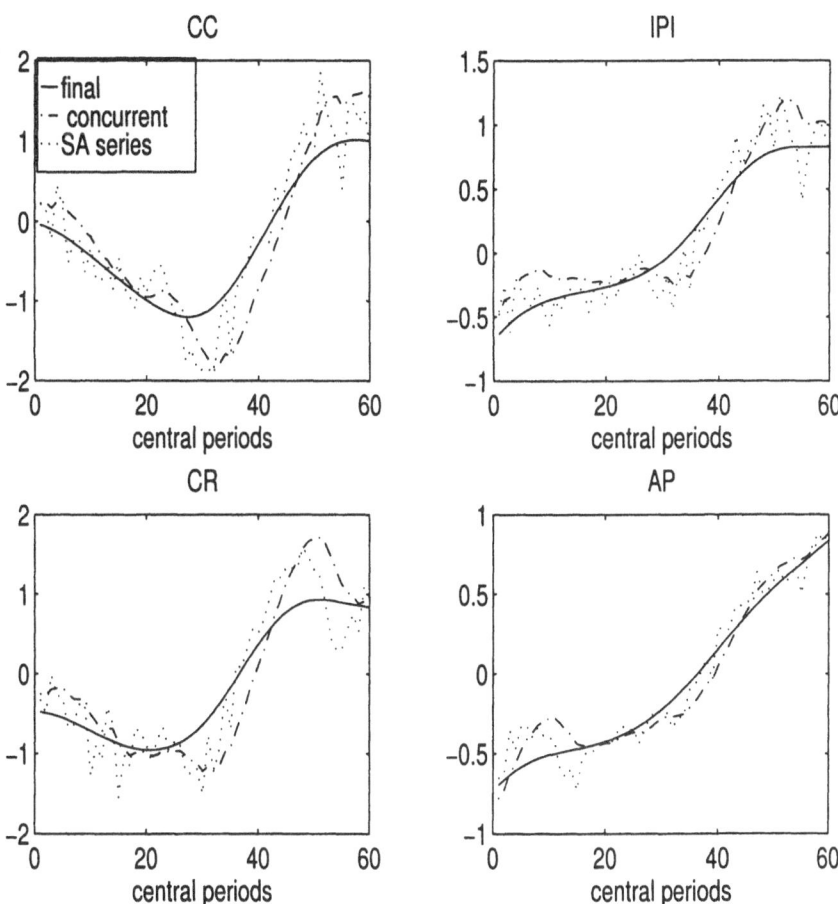

Figure 5.4. Concurrent versus final trend estimator.

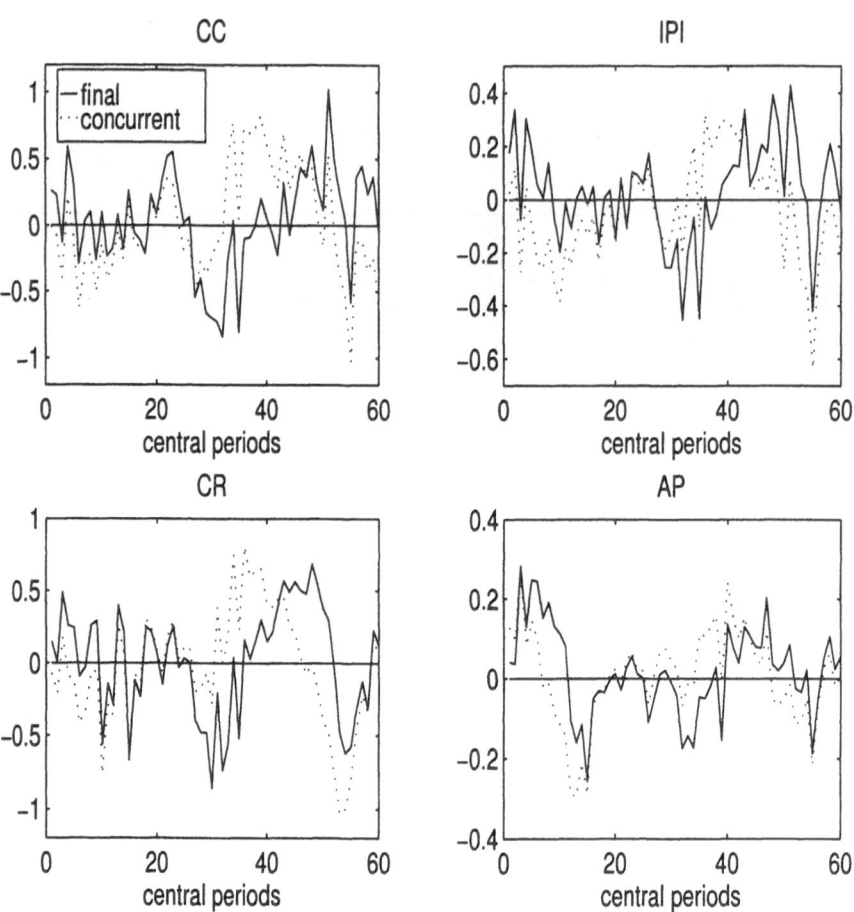

Figure 5.5. Concurrent versus final cycle estimator

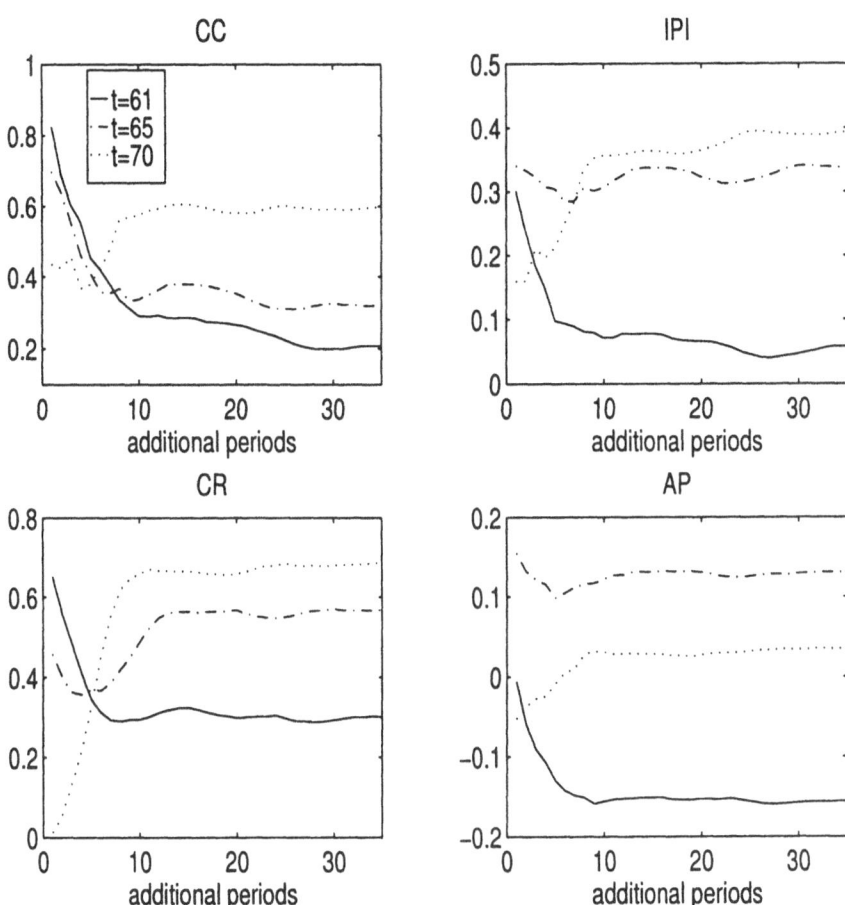

Figure 5.6. Revisions in concurrent estimator.

Table 5.5. Concurrent and Final Cycle Estimator for Three Periods

	Period	t=61	t=65	t=70
CC	Concurrent	.83	.70	.44
	Final	.21	.32	.60
IPI	Concurrent	.30	.34	.16
	Final	.06	.34	.39
CR	Concurrent	.65	.46	.01
	Final	.30	.57	.69
AP	Concurrent	-.01	.16	-.05
	Final	-.16	.13	.04

A point of applied relevance is to assess the imprecision of the estimator of the cycle for recent periods, as measured by the standard error of the revision. Computing the ξ-weights of the filter (5.2), and $Var(r_{t|t})$ as in (5.5), the variance of the revision in any preliminary estimator can be computed as

$$Var(r_{t|t+k}) = Var(r_{t|t}) - \sum_{j=1}^{k}(\xi_j)^2,$$

given that $a_{t+1}, \ldots, a_{t+k}$ have been "observed" at period $t + k$. Insofar as the revision represents a measurement error, its variance can be used to build confidence intervals around the cycle estimator. Figure 5.7 displays the 95% confidence interval for the estimator of the cycle for the four series. Direct inspection shows that, although the estimator converges in two (at most three) years, the estimator for recent periods is unreliable. This fast and large increase in the measurement error of the most recent signals implies that, although straightforward to obtain, forecasts would be close to useless. (In computing revisions, the X11-SA series for the full sample of 104 observations has remained constant. The revisions we have computed are thus those implied solely by the HP filter.)

We come back to the issue of revisions in the next chapter; until then we center our attention on final estimators.

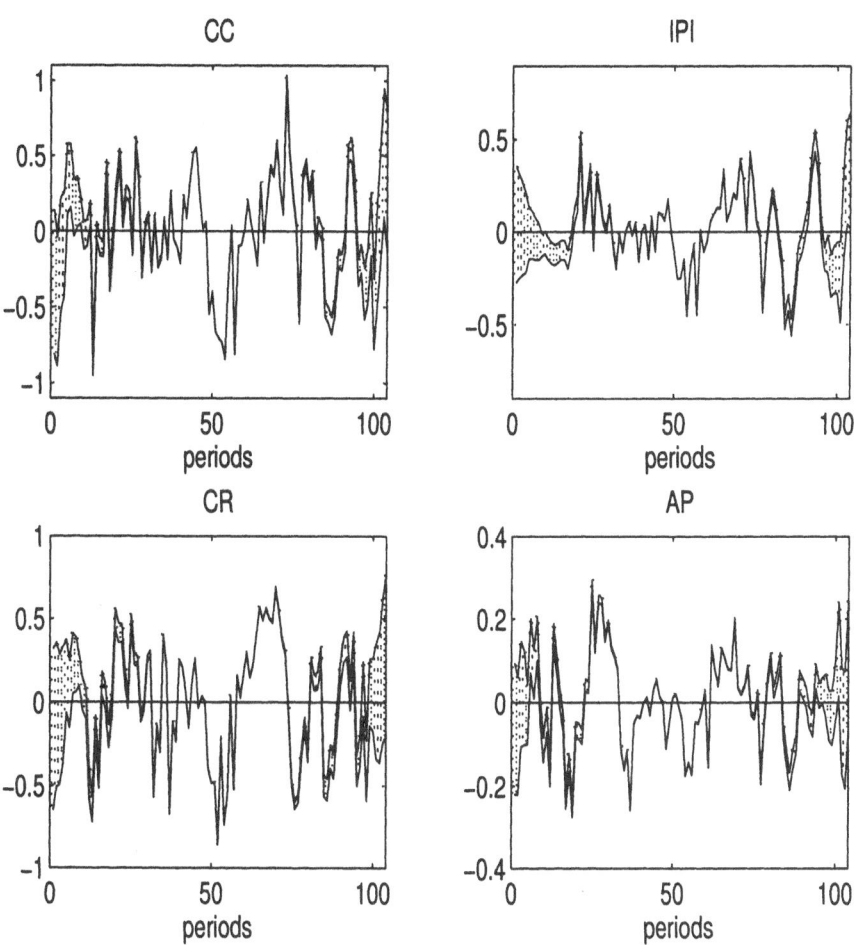

Figure 5.7. 95% Confidence intervals for cycle (based on revisions).

5.2   Spurious Results

The danger of obtaining spurious results induced by HP filtering has been frequently mentioned. To this issue we turn next.

In Section 4.3 we illustrated the basic spuriousness problem. Heuristically, if the filter is designed to remove a nonstationary trend, it is likely to induce a spectral zero for the zero frequency in the detrended series. Likewise, if the filter is designed to remove nonstationary seasonality, it is likely to induce spectral zeros for the seasonal frequencies $\omega = \pi/2$ and $\omega = \pi$ in the SA series. The presence of the three zeros will induce peaks in the spectrum of the filtered series for the cyclical and intraseasonal ranges of frequency. In Section 4.3 we illustrated the effect with a simple model-based filter example. In terms of the X11-HP filter, given by (4.12), the squared gain of the filter $\nu_{HPX}^{c}(B, F)$ is shown in Figure 5.8. It displays zeros for the zero and seasonal frequencies. (In the model-based interpretation of the HP and X11 filters these zeros are implied by the presence of ∇^2 and of S in the autoregressive polynomials for the trend and for the seasonal component models, as shown in Section 4.3.)

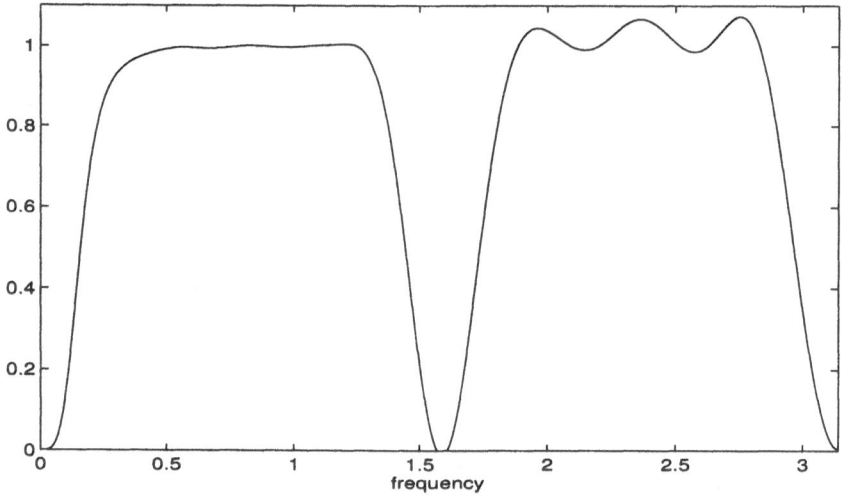

Figure 5.8. Squared gain: convolution of HP and X11 filters.

Assuming a white noise input, the squared gain becomes the spectrum of the estimated cycle. As Figure 5.8 indicates, this spectrum displays two wide peaks, one for a frequency in the range $(0, \pi/2)$, the range of cyclical frequencies, and the other for a frequency in the range $(\pi/2, \pi)$, the range of intraseasonal frequencies. The two-peak structure of the spectrum brings the possibility of obtaining spurious results. It will affect the autocorrelation structure of the series and, due to the common structure, spurious correlations between series may be obtained (in the line of Granger and Newbold (1974)). On the other hand, the first peak may induce a spurious periodic cycle.

5.2.1 *Spurious Crosscorrelation*

We performed a simulation in MATLAB, whereby 10,000 independent random samples of 600 observations each were drawn from an $N(0,1)$ distribution. Each white noise series was filtered through the X11 and HP filters and the last 100 values were selected. Next, 10,000 lag-zero crosscorrelations between two series were sampled (in what follows, all crosscorrelations are lag-zero ones). The average of the absolute value of the crosscorrelation between the white noise input series was .08 (SE = .06), between the seasonally adjusted series, .09 (SE = .06), and between the cycles, .09 (SE = .07). Not much crosscorrelation seems to have been induced by the filter. Table 5.6 presents the first four moments of the distribution of $\hat{\rho}_0$, the crosscorrelation estimator (including the sign) for the original series and the cycle. Figure 5.9 plots the two densities.

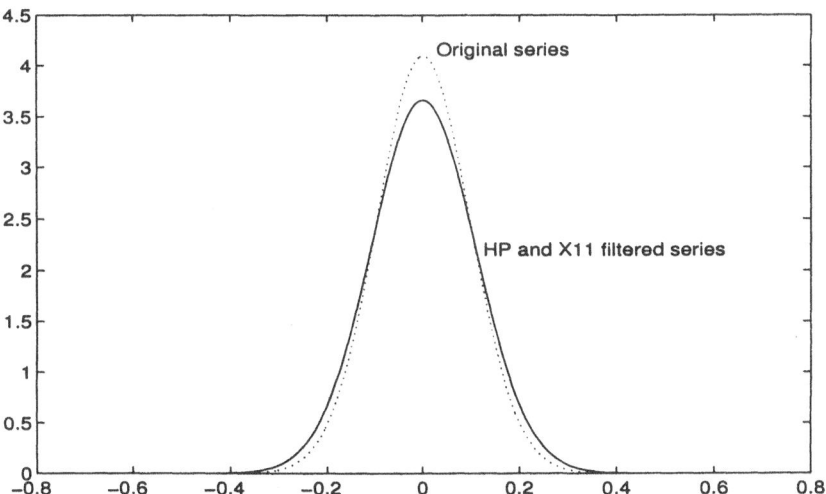

Figure 5.9. Density for correlation coefficient: white noise case.

Table 5.6. Crosscorrelation: Filtered White Noise Case

	Mean	Std. deviation	Skewness	Kurtosis
Original (White Noise)	-.001	.11	-.03	2.9
Cycle	-.001	.11	-.05	2.9

The two distributions are very close and well approximated by $N(0, 1/T)$. Clearly, no spurious significant crosscorrelations have been induced.

When the input follows the random walk model $\nabla x_t = a_t$, using the same simulation as in the previous section, the average over the 10,000 absolute values of the crosscorrelation between the differenced series is .08(.06), and between the differenced SA series, .09 (.07), the same values as before. For the cycle, however, the average increases to .16 (.11), still a small value. Table 5.7 presents the first four moments of $\hat{\rho}_0$ (with sign included) for the original series and for the cycle; the two densities are plotted in Figure 5.10.

Table 5.7. Crosscorrelation: Filtered Random Walk Case

	Mean	Std. Deviation	Skewness	Kurtosis
Original (Random Walk)	-.000	.10	-.04	2.9
Cycle	.000	.19	-.01	2.8

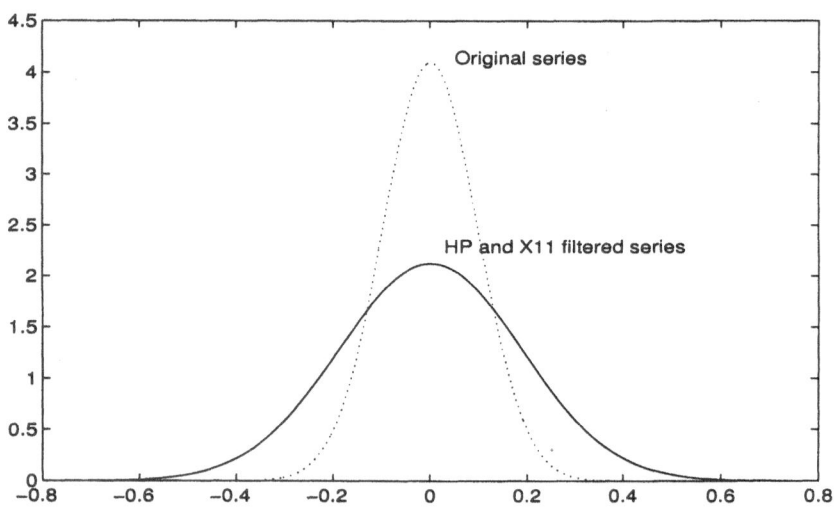

Figure 5.10. Density for correlation coefficient: random walk case.

The zero-mean normality assumption can still be accepted comfortably, but the spread of the distribution of $\hat{\rho}_0$ for the cycle becomes wider. In fact, the proportion of ρ_0 estimates for the cycle that lie outside the 95%

confidence level for the estimate in the stationary original series is 32%. For the random walk case, thus, a spurious crosscorrelation effect can be detected. Altogether, the effect is nevertheless moderate.

A similar simulation was performed for the more complex airline model (3.19), with the parameter values set at $\theta_1 = -.4$ and $\theta_4 = -.6$. Figure 5.11 plots the densities of the crosscorrelation estimator for the stationary transformation of the original and SA series and of the X11-SA and HP detrended series.

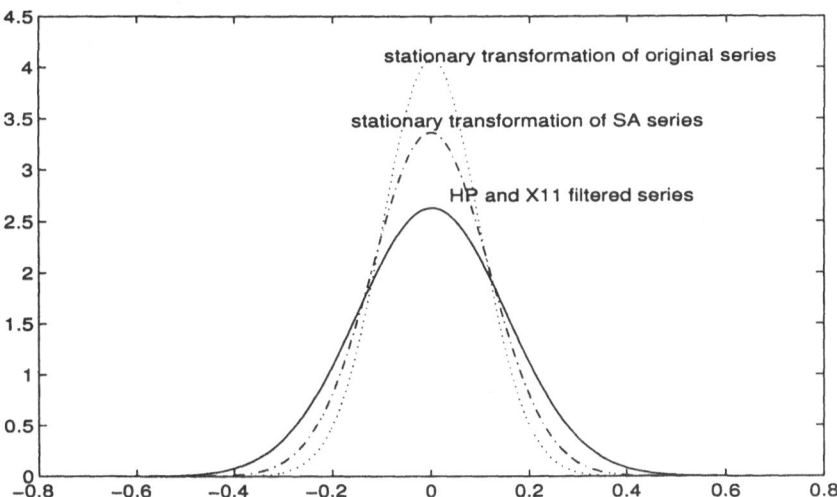

Figure 5.11. Density for correlation coefficient: airline model.

The filter X11 is seen to have virtually no effect while, as before, the HP filter induces a small increase in the spread of the distribution. In summary, from the point of view of spurious crosscorrelation, the HP-X11 filter seems to induce a small amount of spuriousness and hence the detection of relatively large crosscorrelations between cycles obtained with the filtered series is unlikely to be purely spurious. (Although the filter will have some distorting effect on the crosscorrelations when the series are indeed correlated; see Cogley and Nason (1995).)

5.2.2 Spurious Autocorrelation; Calibration

Assume that a theoretical economic model implies that a particular variable should follow a four-year cycle given by the AR(2) process:

$$(1 - 1.293B + .490B^2)c_t = a_{ct} \tag{5.6}$$

with a_{ct} a white noise innovation, and variance $V_c = 1$. Assume that a large number of simulations of the model yield in fact an ACF for the variable equal to the theoretical ACF of (5.6), shown in the second column of Table 5.8. The basic idea behind calibration is to validate the economic model by comparing the previous ACF with the one implied by the observed economic variable. To compute the latter, the nonstationary trend and seasonal component need to be removed. (Seasonality and often the trend are typically excluded from the theoretical economic model.)

Assume the observed series x_t is given by $c_t + p_t + s_t$, where c_t is generated precisely by the cycle given by (5.6), contaminated by a random walk trend (p_t), and a seasonal component (s_t), as in

$$\nabla p_t = a_{pt}$$
$$S s_t = a_{st}$$

with a_{ct}, a_{pt}, and a_{st} mutually orthogonal innovations, with variances V_c, V_p, and V_s.

Table 5.8 Theoretical ACF of the Component Model and Its Estimators

Lag-k ACF	True Component	X11-HP Filtered Component			MMSE Estimator
		$V_p = .1,$ $V_s = .1$	$V_p = .1,$ $V_s = 1$	$V_p = 1$ $V_s = 1$	$V_p = .1,$ $V_s = 1$
$k=1$	.87	.71	.19	.37	.83
$k=2$	.63	.44	.22	.30	.43
$k=3$	.39	.10	-.06	.00	-.02
$k=4$	.20	-.05	.22	.18	-.35
$k=5$	.06	-.25	-.23	-.15	-.45
$k=6$	-.01	-.30	-.19	-.16	-.43
$k=7$	-.05	-.34	-.27	-.26	-.31
$k=8$	-.06	-.27	-.01	-.07	-.20
$k=9$	-.05	-.25	-.18	-.20	-.10
$k=10$	-.04	-.19	-.12	-.16	-.04
$k=11$	-.02	-.16	-.17	-.20	-.00
$k=12$	-.01	-.09	.06	-.03	-.02
$k=13$	-.00	-.08	-.07	-.12	-.03
$k=14$	-.00	-.05	-.03	-.08	-.03
$k=15$	-.03	-.04	-.11	-.13	-.02
$k=16$	.00	.01	.13	.05	.01

Seasonally adjusting (with X11) and detrending (with the HP filter) the "observed" series, the estimator of the cycle is obtained. Its variance and ACF (the observed moments in the calibration comparison) are straightforward to derive analytically; they are given in the third, fourth, and fifth columns of Table 5.8 for the three cases $V_p = V_s = .1$; $V_p = .1, V_s = 1$; and $V_p = V_s = 1$. Comparing these three columns with the second, the ACF of the cycle contained in the series and of the one obtained by filtering will differ considerably. Although the theoretical model is perfectly correct, the second moments obtained from the observed series would seem to indicate the contrary.

The distortion that seasonal adjustment and detrending induce in the second moments of the series is a general property that also occurs when the components are estimated as MMSE estimators in a model-based approach. The ACF of the cycle obtained in this case is given in the sixth row of Table 5.8; the distortion induced by MMSE estimation is considerably smaller than that induced by HP-X11 filtering.

Calibration of models using filtered series seems, thus, an unreliable procedure. If the theoretical economic model is correct, then calibration should not look for similarities between the ACF of the theoretical model and of the empirical filtered series. It should compare instead the empirical moments with the theoretical ones that include the effect of filtering the data. Performing this comparison, however, requires incorporating in some way into the model trend and seasonality (in its simplest way, as unobserved components; see Harvey and Scott (1994)).

5.2.3   Spurious Periodic Cycle

As mentioned at the beginning of this section, the HP filter has often been accused of inducing spurious cycles in the filtered series. To analyze the effects of a filter, attention usually centers on its gain function. But this function only tells part of the story. It is seen in equation (2.37) that the spectrum of the cycle obtained is the convolution of the squared gain of the filter and the spectrum of the observed series. The squared gain tells us which variation would be passed on to the cycle (according to the different frequencies). To see which variation is actually passed, we need to know which ones are present in the series; this is what the series spectrum tells us. Thus a more informative function to analyze the properties of the cycle obtained is its spectrum. Accordingly, in order to analyze the spuriousness issue, we mostly look at the spectrum of the estimated cycle.

(1) White noise input

A priori, one can think of capturing the two spectral peaks of Figure 5.8 through an AR(4) model with two pairs of complex conjugate roots. We performed the same simulation of Section 5.2.1 (10,000 random samples from an $N(0,1)$ distribution were filtered through the X11 and HP filters, and the last 100 values were selected from each series). Then, an AR(4) model was fit to the filtered series (i.e., to the "cycle"). Averaging the AR parameter estimates yields very approximately the model

$$(1 + .31B^4)c_t = a_t, \tag{5.7}$$

and a test for the significance of the AR(4) regression yields an average F value of 3.6 (critical value 2.7). Figure 5.12 plots the spectrum of model

(5.7) and it is clearly seen how the peaks of the AR approximation attempt to capture the peaks of the spectrum of the filtered white noise. It is also seen that the AR format does not permit a good approximation, reflecting the fact that the invertible AR model cannot approximate well the spectral zeros of the noninvertible cycle. The two spectral AR peaks correspond to two pairs of complex conjugate roots for the AR polynomial. The average value of the modulus and period for the two roots are given in Table 5.9, where the standard errors are given in parentheses. (It is easily seen that factorization of the AR(4) polynomial in (5.7) yields nearly identical roots.)

Table 5.9. AR Roots: Filtered White Noise

	Average Modulus	Average Period
First Root	.75 (.06)	7.95(.79)
Second Root	.74 (.06)	2.68(.08)

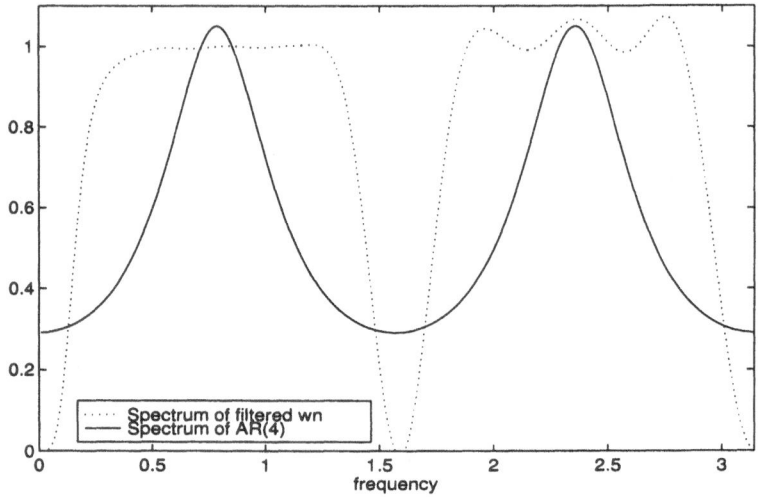

Figure 5.12. Spectrum of AR(4) for the white noise case.

In summary, the HP-X11 procedure is likely to induce spurious cycles in a white noise series. If captured with an AR model, these cycles are of similar amplitude and clearly significant. One of the components has a 2-year period; the other one has a period of 2.7 quarters. The sum of the two produces a highly erratic series and it is difficult to link this behavior to the concept of a business cycle. Figure 5.13 presents an example where a white noise series is decomposed, with the HP-X11 procedure, into a trend, a seasonal component, and a cycle. Insofar as we do not wish to remove trends nor seasonals from white noise, the HP-X11 decomposition would be spurious, purely filter-induced. The spurious trend and seasonality removed from the series are nevertheless moderate. Given that application to white noise simply repeats the filter features, the example serves to illustrate the pure filter effects. However it is unlikely that an analyst would use X11-HP filtering to find the cyclical component in an obviously stationary series. We move on to more relevant examples.

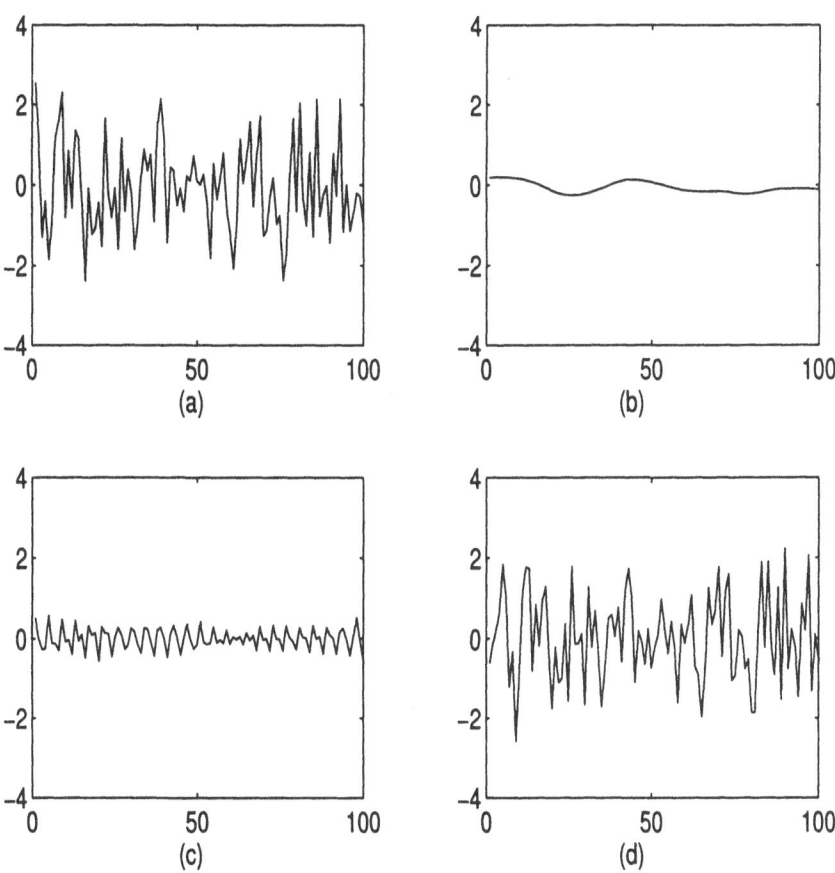

Figure 5.13. Decomposition of a white noise series: (a) original white noise series; (b) trend component; (c) seasonal component; (d) cycle component.

(2) Random walk input

The previous discussion illustrated the filter effect and it is of interest to see how this effect interacts with input that displays some trend structure. We consider the simplest case of the random walk model,

$$\nabla x_t = a_t; \tag{5.8}$$

as before, the cycle is estimated through (4.24).

From (2.37), if $\tilde{\nu}_{HPX}^c(\omega)$ is the Fourier transform of $\nu_{HPX}^c(B, F)$, the spectrum of the estimator of the cycle is given by,

$$\hat{g}_{HPX}^c(\omega) = [\tilde{\nu}_{HPX}^c(\omega)]^2 g_x(\omega), \tag{5.9}$$

where $g_x(\omega)$ is the spectrum of x_t. Figure 5.14 plots the spectra of the series (dotted line) and of the cycle (continuous line). The difference with respect to the cycle obtained with white noise (Figure 5.12) is remarkable. The peak for the high frequency is hardly noticeable, while the peak for the frequency in the cyclical range is associated with a longer period of 31 to 32 quarters, or, approximately, eight years.

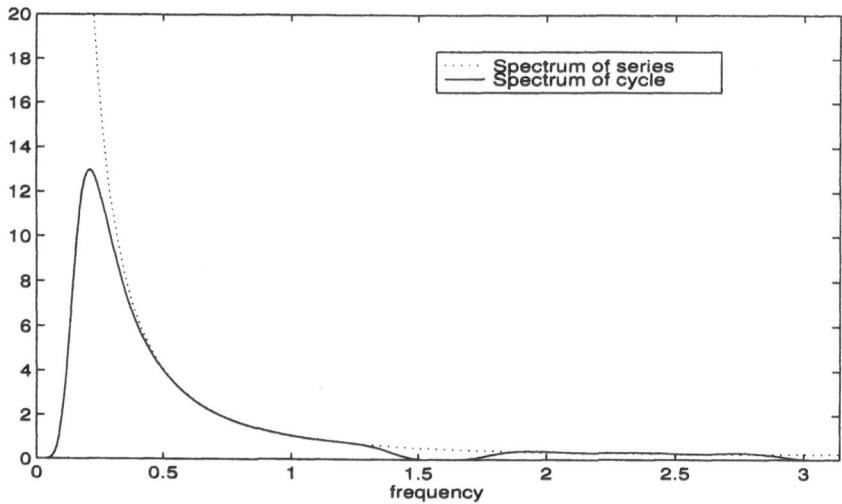

Figure 5.14. Spectrum of cycle component in a random walk.

Performing the same simulation as before, an AR(4) model was fit to 10,000 generated random walks of 100 observations each, filtered through the X11 and HP filters, and the average F-test was equal to 38.35, overwhelmingly significant. For the random walk series, the HP-X11 filter induced a cycle dominated by an eight-year period, and hence more in line with the frequencies of interest to business cycle analysts. Figure 5.16 presents an example of a random walk decomposed by the X11 and HP filters into a trend, a seasonal component, and a cycle. Although the trend and cycle are, as before, moderately small, by its own definition, does it make sense to see a random walk as generated by a trend, a seasonal component, and an eight-year cycle? Is it not rather a case of "overreading" the data? The answer to this question is, as we show, not quite so obvious.

(3) Spectral characteristics of the cycle

The two previous examples show that the cycle obtained with X11-HP filtering displays a stochastic cyclical structure, with spectrum given by the general expression (5.9). This spectrum depends on the ARIMA model followed by the observed series, and on the λ-parameter of the HP filter.

To look at the effect of the model, we set $\lambda = 1600$. Figure 5.15 compares the cycles obtained when the series follows the IMA(1,1) model $\nabla x_t = (1 + \theta B)a_t$, with $V_a = 1$, for a range of values for θ. In all cases, the period of the cycle (i.e., the period associated with its spectral peak) is approximately constant, and very close to eight years. The amplitude of the cycle varies, adapting to the width of the spectral peak for $\omega = 0$ in the series model, which is determined by the parameter θ.

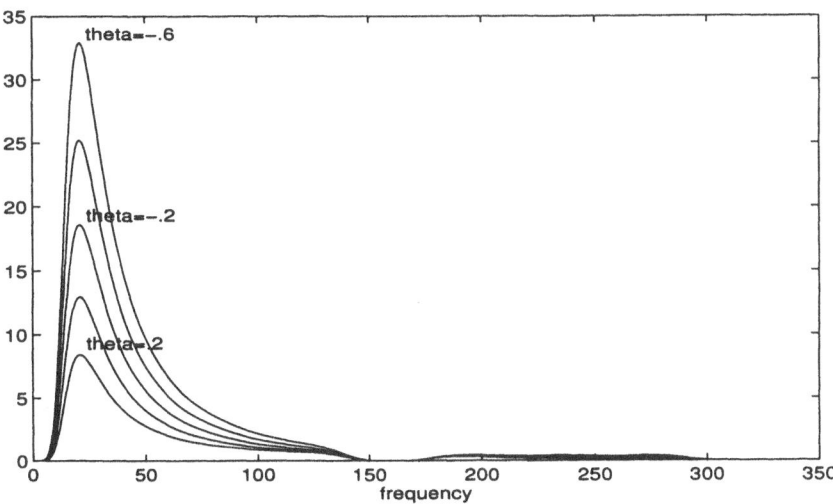

Figure 5.15. Spectrum of cycle in IMA(1 , 1) as a function of θ.

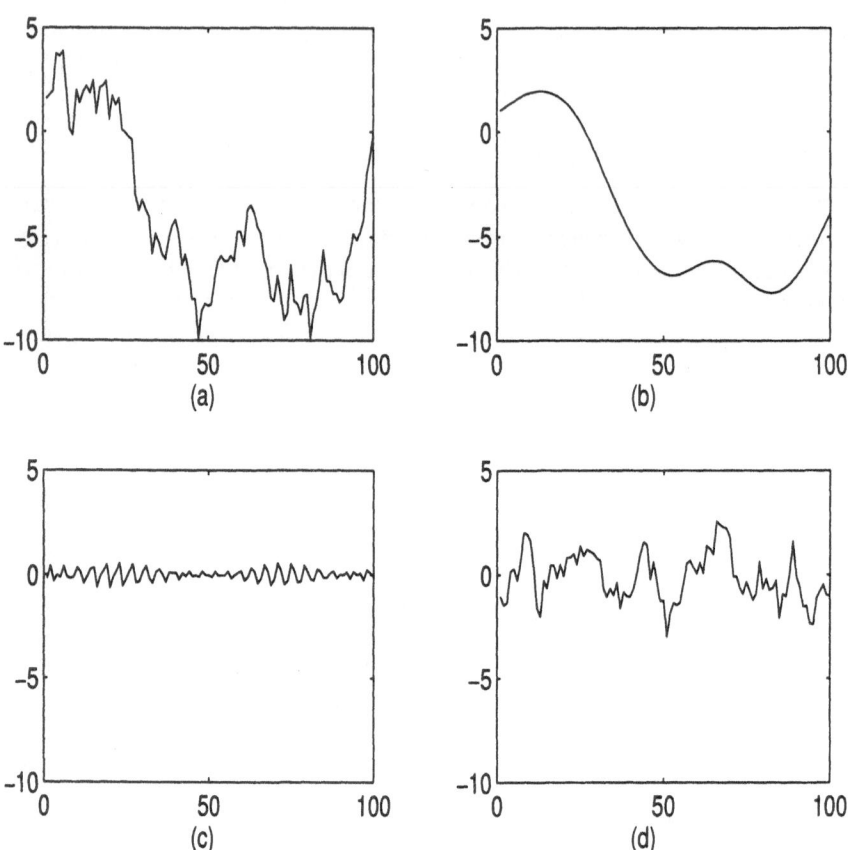

Figure 5.16. Decomposition of a random walk series: (a) original random walk series; (b) trend component; (c) seasonal component; (d) cycle component;

The relative constancy of the period with respect to the model parameter is also shown in Table 5.10 for a MA(1) and an IMA(2,1) models. What the table seems to indicate is that, for a fixed value of λ, the period of the cycle is determined fundamentally by the order of integration of the series, rather than by the model parameters. As the order of integration increases, so does the period of the cycle.

Table 5.10. Period of Cycle (in Years)

Theta	0	-.3	-.6
MA(1)	2	3	3.2
IMA(1,1)	7.9	7.9	7.9
IMA(2,1)	10.5	10.5	10.5

When the HP filter is applied to an X11-SA series, a similar effect is seen to occur. For the airline model (3.19) we computed the period associated with the spectral peak of the cycle for the range $-.9 < \theta_1 < .5$ and $-.9 < \theta_4 < 0$, and in all cases the period was equal to approximately 10 years. For fixed parameter λ, three conclusions emerge concerning the period of the cycle spectral peak.

- Given the type of ARIMA model for the series, the associated cyclical period becomes roughly fixed.

- The period seems to be mostly determined by the order of integration at the zero frequency; the stationary part of the model has little influence.

- For most actual time series containing a trend ($d = 1$ or 2), the standard value of $\lambda = 1600$ implies a period between 8 and 10 years.

As for the influence of the parameter λ of the HP filter on the cycle obtained, fixing the series model to that of a random walk, it is found that $g(x) = [2(1 - \cos \omega)]^{-1} V_a$. Considering (5.9), with $\tilde{\nu}_{HPX}^c(\omega)$ replaced by $\tilde{\nu}_{HP}^c(\omega)$, and (4.25), the spectrum of the HP-filtered cycle is equal to

$$g_{HP}^c(\omega) = \frac{8(1 - \cos \omega)^3}{[\lambda^{-1} + 4(1 - \cos \omega)^2]^2} V_a. \tag{5.10}$$

It is straightforward to find that, within the interval $0 \leq \omega \leq \pi$, (5.10) attains a single maximum at

$$\lambda = \frac{3}{4(1 - \cos \omega)^2}.$$

For the range of frequencies associated with periods between 2 and 25 years, this maximum is represented in Figure 5.17. It can be seen that the convolution with X11 has little effect on the period of the cycle peak (in fact the two figures would be indistinguishable). This was to be expected, given that, for the range of frequencies where the spectral peak is located, the gain of the X11 filter is close to 1. The values of the parameter λ, and the associated period of the cycle (in years), are displayed in the second and fourth columns of Table 5.11. The relationship between λ and the period of the cycle spectral peak is seen to be highly nonlinear. When λ is small (and cycles are short), small increases in λ affect very strongly the period of the cycle; for long cycles, very large values of λ need to be used.

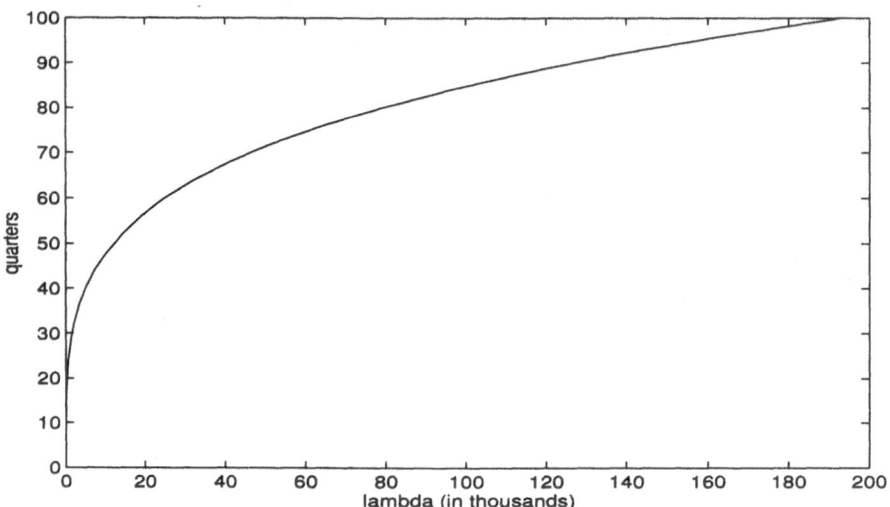

Figure 5.17. Period of cycle as a function of λ.

Table 5.11. Values of λ for Different Cycles (Period in Years)

Period (in Years)	λ (Approx.)	Period (in Years)	λ
2	9	14	18970
3	42	15	24992
4	129	16	32346
5	313	17	41215
6	646	18	51794
7	1193	19	64291
8	2031	20	78924
9	3250	21	95923
10	4948	22	115532
11	7239	23	138004
12	10247	24	163605
13	14108	25	192614

The effect of λ on the cycle is illustrated in Figure 5.18, which compares the spectra of the cycles obtained with $\lambda = 1600$ and $\lambda = 25,000$ for the same random walk series (the periods associated with the spectral peaks are about 8 and 15 years, respectively). The figure shows that the longer period implies a stochastic cycle with a much larger variance, although, in relative terms, it is more concentrated around its peak. The estimators of the trend and of the cycle for the two λ values are compared in Figures 5.19 and 5.20, respectively. The difference between the two trends is seen to consist mostly of a cycle with a relatively long period. Comparison of the cycles shows that the short-term profile of the cycle is basically unaffected, and the main effect is a "pulling away" from the zero line, which allows for longer cycles. (An interesting example of the effect of λ in a particular application can be found in Dolado et al. (1993).)

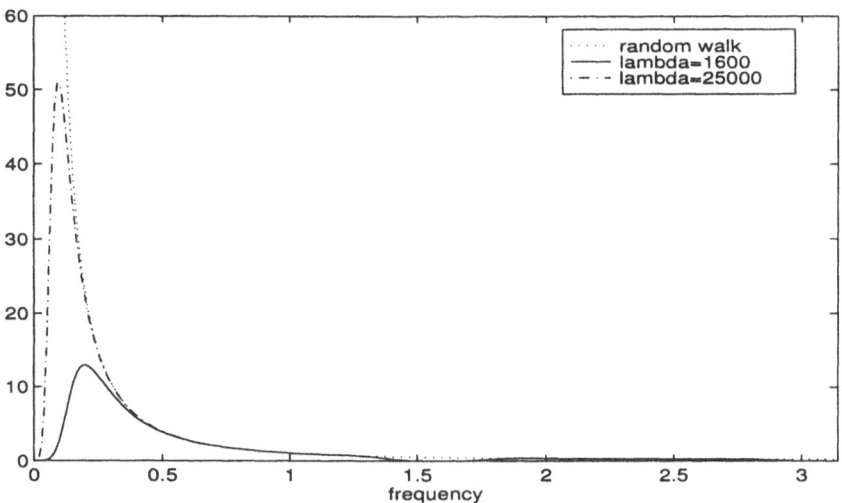

Figure 5.18 Spectrum of a cycle in a random walk.

As a consequence, the use of the X11-HP filter (or simply the HP filter) to measure the cycle implies an a priori choice of the cycle period. Before using the HP filter to estimate a cycle, the analyst should decide the length of the period around which he wishes to measure cyclical activity. Then, he can choose the appropriate value of λ. To some extent, this may be reasonable. For example, a business cycle analyst involved in policy making may be interested in using 8-or 10-year cycles; an economic historian looking at several centuries, may be interested in spreading activity over longer periods.

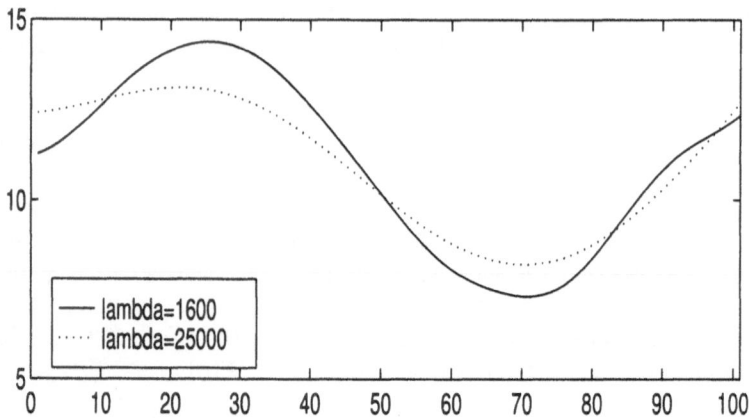

Figure 5.19. Estimated trends.

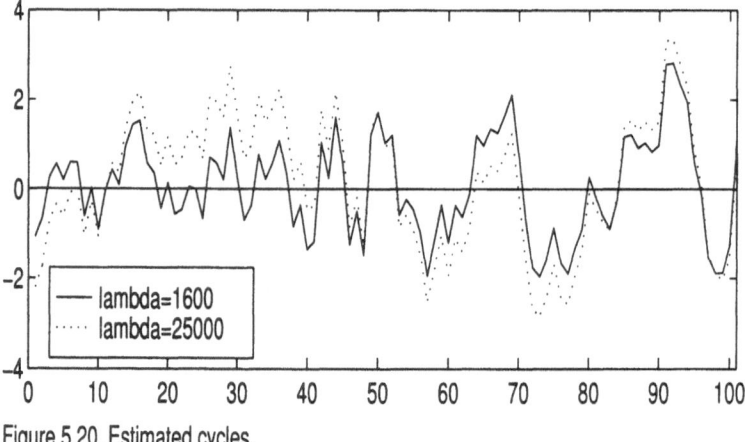

Figure 5.20. Estimated cycles.

5.3 Noisy Cyclical Signal

Whether the HP filter is used by itself or together with X11, the squared gain of the filter to obtain the cycle (Figures 4.3(a) or 5.8) shows that the filter will pass to the cycle high-frequency variation, such as the one associated with intraseasonal frequencies, as well as a large amount of the noise component present in the observed series. These short-term, highly transitory variations, that correspond to frequencies of no cyclical interest, will act as noise contaminating the cyclical signal. The fact that HP filtering may yield a cycle containing contributions from noncyclical frequencies was already pointed out by Nelson and Plosser (1982) and is one of the drawbacks addressed by Baxter and King (1999) for proposing their alternative.

Figure 5.3 illustrated well the erraticity of the cyclical signal. It is seen, for example, that the cycle for the CC series crosses the horizontal zero line 35 times; the cycles for the IPI, CR, and AP series cross the zero line 25, 27, and 29 times, respectively, These are unreasonably high numbers given the length of the period considered (104 quarters), and the noise present in the cyclical signal makes a proper reading difficult. The outcome of the noise contamination is an artificial increase in the variance of the cycle component, that induces a manic-depressive behavior in short-term economic monitoring, and confuses identification of turning points, even when historical estimates are used.

6.

Improving the Hodrick–Prescott Filter

In Chapter 5 we saw that the filter implies large revisions for recent periods (roughly, for the last two years). The imprecision in the cycle estimator for the last quarters implies, in turn, a poor performance in early detection of turning points. Furthermore, as was just mentioned, direct inspection of Figure 5.3 shows another limitation of the HP filter: the cyclical signal it provides seems rather uninformative. Seasonal variation has been removed, but a large amount of noise remains in the signal, making its reading and the dating of turning points difficult. In the next two sections, we proceed to show how these two shortcomings can be reduced with some relatively simple modifications.

6.1   Reducing Revisions

As is the case with two-sided fixed filters, estimation of the cycle for the end periods of the series by the HP filter implies an ad hoc, somewhat abrupt, discontinuous truncation of the filter. As mentioned at the end of Section 4.4, in terms of the model-based derivation, the truncation is equivalent to the assumption that model (4.20) and (4.22) is always the model that generates forecasts and backcasts to extend the series at both endpoints. The assumption is in general false, and optimal forecasts and backcasts (obtained with the appropriate ARIMA model for the series) can be used instead to improve the filter extension. This idea is the same as the one behind the X11-ARIMA modification of the X11 filter (see Dagum (1980)) and the HP filter applied to the series extended with ARIMA forecasts is referred to as the Hodrick–Prescott ARIMA (HPA) filter. The poor performance of the HP filter at the end of the series often has been pointed out by applied business-cycle analysts (see, e.g., Apel et al. (1996) and Baxter and King (1999)) and application of the filter to series extended with forecasts is, on occasion, recommended in practice (see EU Commission (1995)).

Reasoning as in Section 3.4, for any positive integer k, write the historical estimator of the cycle for period t as

$$\hat{c}_t = \nu_{HP}^c(B, F)x_t = \sum_{j=0}^{\infty} \nu_{k-j} x_{t+k-j} + \sum_{j=1}^{\infty} \nu_{k+j} x_{t+k+j} \qquad (6.1)$$

($\nu_{-j} = \nu_j$), and assume a series long enough so as to ignore starting values. Because the preliminary estimator $\hat{c}_{t|t+k}$ is a projection onto a subset of the set onto which $\hat{c}_t$ is projected, it follows that

$$\hat{c}_{t|t+k} = E_{t+k}(c_t) = E_{t+k}(\hat{c}_t),$$

or

$$\hat{c}_{t|t+k} = \sum_{j=0}^{\infty} \nu_{k-j} x_{t+k-j} + \sum_{j=1}^{\infty} \nu_{k+j} E_{t+k}(x_{t+k+j}), \qquad (6.2)$$

which expresses the preliminary estimator of the cycle for period t, when observations are available up to and including x_{t+k}, as a function of the series extended with forecasts. Subtracting (6.2) from (6.1), the revision in $\hat{c}_{t|t+k}$ is equal to the convergent sum

$$r_{t|t+k} = \sum_{j=1}^{\infty} \nu_{j+k} e_{t+k+j|t+k},$$

where $e_{t+k+j|t+k}$ denotes the forecast error associated with forecasting x_{t+k+j} at time $t + k$. It follows that if these forecast errors are reduced, revisions should decrease, and early detection of turning points should improve.

To check these results, and to get an idea of the improvement that can be expected from the use of the HPA versus the HP filter we performed a simulation exercise. First, we considered the IMA(1,1) model for different values of the θ-parameter. Then, we considered the ARIMA(2,1,1) model, where the AR(2) polynomial was given by $(1 - .16B + .35B^2)$. This polynomial is the one found in Jenkins (1975) for the mink-muskrat Canadian data, and contains a cycle of period 4.4. The AR(2) structure therefore produced an increase in the number of turning points. Again, different values of the θ-parameter were considered. A total of 14,000 series of length 100 each were simulated, and for each series the HP filter was compared to the HPA one extended with 16 ARIMA forecasts and backcasts. Table 6.1 compares the variances of the revision in the concurrent estimator and in the estimator revised after 1, 2, 3, and 4 more years of data were added. It is seen that, in all 70 cases (2 models × 7 parameters × 5 revision period lengths), the HPA filter reduces the revisions considerably. This is particularly noticeable for the ARIMA(2,1,1) model, where the use of the standard HP filter more than triples the revision error variance.

Table 6.1. Variance of the Revision in Estimator. Values are Multiplied by 100. Model 1 Is IMA(1,1) and Model 2 Is ARIMA(2,1,1)

Model	Conc.		1 Year Rv.		2 Year Rv.		3 Year Rv.		4 Year Rv.	
	HPA	HP	HPA	HP	HPA	HP	HPA	HP	HPA	HP
1										
$\theta = -.8$	.31	.41	.08	.10	.02	.03	.01	.01	.00	.01
$\theta = -.5$	.94	1.34	.24	.33	.07	.11	.03	.06	.02	.04
$\theta = -.3$	1.58	2.54	.39	.63	.11	.19	.06	.12	.04	.08
$\theta = 0$	2.86	4.84	.71	1.18	.21	.39	.12	.24	.08	.15
$\theta = .3$	4.51	8.29	1.11	2.03	.32	.64	.19	.38	.13	.24
$\theta = .5$	5.44	11.02	1.40	2.62	.44	.86	.26	.50	.17	.32
$\theta = .8$	7.33	14.89	1.87	3.70	.60	1.17	.34	.70	.22	.46
2										
$\theta = -.8$	.12	.55	.05	.14	.01	.03	.00	.01	.00	.01
$\theta = -.5$	.41	1.27	.15	.33	.04	.09	.01	.04	.00	.03
$\theta = -.3$	.74	2.23	.25	.56	.06	.15	.02	.08	.01	.05
$\theta = 0$	1.35	4.26	.44	1.09	.12	.29	.03	.15	.02	.10
$\theta = .3$	2.06	7.00	.71	1.77	.18	.46	.05	.25	.03	.18
$\theta = .5$	2.75	9.68	.95	2.44	.23	.64	.06	.33	.03	.22
$\theta = .8$	3.70	12.95	1.20	3.25	.31	.88	.09	.47	.05	.33

As for detection of turning points, we use the following simple criterion (along the lines of method B discussed in Boldin (1994)): a turning point is the first of at least two successive periods of negative/positive growth. Table 6.2 compares the performance of the HP and HPA filters in the first and last eight observations of the simulated series, both in terms of the mean number of turning points that are dated in the original series and missed by the filtered one, and in terms of the mean number of turning points detected in the filtered series but not present in the original one ("peaks" and "throughs" are considered separately). Of the 56 comparisons, in 53 cases the gain from using the HPA filter is substantial.

Tables 6.3 and 6.4 compare the performance of the two filters when all observations in each series are considered. Table 6.3 compares the performance in detecting turning points present in the original series, and Table 6.4 looks at the false turning points indicated by the filtered series (and not present in the original one). For both tables, F_0 is the relative frequency of cases in which the two filters coincide, F_1 denotes the relative frequency of cases in which HPA performs better, while F_{-1} denotes the relative frequency of cases in which HP performs better. Both tables show that the HPA filter performs (in all 56 cases) remarkably better.

Table 6.2. Mean Number of Turning Points (First and Last 8 Observations). P: Peaks, T: Throughs

IMA(1,1)	Original		Missed				False Alarms			
	P	T	P		T		P		T	
			HPA	HP	HPA	HP	HPA	HP	HPA	HP
$\theta = -.8$	1.4	1.50	.10	.16	.10	.15	.18	.19	.19	.20
$\theta = -.5$	1.51	1.52	.18	.22	.19	.23	.29	.31	.26	.28
$\theta = -.3$	1.52	1.52	.22	.28	.24	.30	.36	.37	.38	.39
$\theta = 0$	1.62	1.59	.23	.32	.25	.37	.49	.51	.47	.49
$\theta = .3$	1.69	1.72	.29	.40	.29	.41	.49	.59	.55	.63
$\theta = .5$	1.77	1.79	.32	.46	.29	.41	.54	.68	.51	.63
$\theta = .8$	1.83	1.86	.30	.43	.34	.48	.52	.65	.54	.69
ARIMA(2,1,1)										
$\theta = -.8$	1.78	1.79	.05	.13	.05	.13	.21	.18	.19	.18
$\theta = -.5$	1.78	1.77	.09	.19	.11	.19	.25	.24	.26	.26
$\theta = -.3$	1.84	1.77	.12	.23	.10	.20	.25	.31	.25	.30
$\theta = 0$	1.87	1.84	.13	.25	.16	.27	.27	.38	.26	.36
$\theta = .3$	1.93	1.89	.17	.27	.20	.31	.25	.40	.25	.41
$\theta = .5$	1.94	1.96	.18	.31	.17	.32	.24	.40	.25	.42
$\theta = .8$	2.02	1.95	.18	.32	.16	.31	.22	.38	.24	.41

Table 6.3. Relative Performance of HP vs. HPA: Captured Turning Points. $F_0 =$ Relative Frequency of Cases in Which the Two Filters Coincide; $F_1 =$ Relative Frequency of Cases in Which HPA Performs Better; $F_{-1} =$ Relative Frequency of Cases in Which HP Performs Better

Model	Capt. Peaks			Capt. Throughs		
	F_1	F_0	F_{-1}	F_1	F_0	F_{-1}
IMA(1,1)						
$\theta = -.8$	.07	.92	.01	.07	.93	.00
$\theta = -.5$	.07	.91	.02	.07	.89	.04
$\theta = -.3$	.11	.84	.05	.10	.86	.04
$\theta = 0$	.14	.80	.06	.16	.80	.04
$\theta = .3$	.17	.78	.05	.17	.78	.05
$\theta = .5$	.18	.76	.06	.19	.74	.07
$\theta = .8$	.18	.76	.06	.20	.75	.05
ARIMA(2,1,1)						
$\theta = -.8$	.11	.88	.01	.11	.88	.01
$\theta = -.5$	.14	.84	.02	.13	.84	.03
$\theta = -.3$	.17	.79	.04	.16	.80	.04
$\theta = 0$	.18	.77	.04	.18	.77	.05
$\theta = .3$	.18	.77	.05	.20	.73	.07
$\theta = .5$	.21	.73	.06	.23	.71	.06
$\theta = .8$	.21	.74	.05	.22	.73	.05

Table 6.4 Relative Performance of HP vs. HPA: Spurious Turning Points. $F_0 =$ Relative Frequency of Cases in Which the Two Filters Coincide; $F_1 =$ Relative Frequency of Cases in Which HPA Performs Better; $F_{-1} =$ Relative Frequency of Cases in Which HP Performs Better

Model	False Peaks			False Throughs		
	F_1	F_0	F_{-1}	F_1	F_0	F_{-1}
IMA(1,1)						
$\theta = -.8$	.06	.91	.03	.05	.91	.04
$\theta = -.5$	.09	.85	.06	.10	.84	.06
$\theta = -.3$	.12	.81	.07	.13	.78	.09
$\theta = 0$	.15	.74	.11	.15	.74	.11
$\theta = .3$	.22	.68	.10	.19	.71	.10
$\theta = .5$	.23	.68	.09	.20	.70	.10
$\theta = .8$	.25	.64	.11	.24	.65	.11
ARIMA(2,1,1)						
$\theta = -.8$	.07	.86	.07	.06	.84	.10
$\theta = -.5$	.10	.80	.10	.10	.81	.09
$\theta = -.3$	.15	.77	.08	.14	.78	.08
$\theta = 0$	.19	.74	.07	.19	.75	.06
$\theta = .3$	.28	.66	.06	.25	.67	.08
$\theta = .5$	.23	.70	.07	.23	.70	.07
$\theta = .8$	.28	.66	.06	.25	.69	.06

In summary, the results of the simulation exercise strongly suggest that applying the HP filter to the series extended at both ends with appropriate ARIMA forecasts and backcasts is likely to provide a more precise cycle estimator for recent periods, that requires considerably smaller revisions, and improves thereby detection of turning points.

6.2 Improving the Cyclical Signal

Concerning seasonality, its removal implies the removal of the spectral peaks associated with seasonal frequencies. Since the width of this peak varies across series, fixed filters such as X11 may ove-r or underestimate seasonality. Having obtained an ARIMA model for the series, one could use, instead of X11, an ARIMA-model-based type of adjustment, following the approach of Section 3.5. We use the SEATS programto seasonally adjust the four series of the example of Section 5.1.

For the airline model (3.19), appropriate for the four series, the AMB method decomposes the series x_t as in

$$x_t = n_t + s_t, \tag{6.3}$$

where n_t denotes the SA series and s_t the seasonal component, which follow models of the type

$$\nabla^2 n_t = \theta_n(B)a_{nt}, \tag{6.4}$$
$$S s_t = \theta_s(B)a_{st}, \tag{6.5}$$

where $\theta_n(B)$ and $\theta_s(B)$ are of orders 2 and 3, respectively. Given the observed series $[x_1, \ldots, x_T]$, the estimator of n_t is the conditional expectation $\hat{n}_{t|T} = E(n_t \mid x_1, \ldots, x_T)$. If $\theta(B) = (1 + \theta_1 B)(1 + \theta_4 B^4)$, applying (3.27) to this particular model, the historical estimator, valid for the central years of the series, is given by the expression

$$\hat{n}_t = \left[k_n \frac{\theta_n(B)S}{\theta(B)} \frac{\theta_n(F)\bar{S}}{\theta(F)} \right] x_t. \tag{6.6}$$

The expression in brackets is the WK filter and it avoids over-underestimation by adjusting itself to the width of the spectral peaks present in the series (see, e.g., Maravall (1995)).

Figure 6.1 compares the cycles obtained by applying the HP filter to the AMB and X11-SA series, and Figure 6.2 exhibits the spectra of the two cycles for the four series. It is seen that the estimates of the cycle produced using the two SA series are close, and no improvement results from applying the AMB method: turning points are basically unchanged, and the cyclical signal remains very noisy. (Figure 6.2 illustrates one of the dangers of fixed filters, namely, overestimation of seasonality for the case of the CC series: seasonality is very stable and consequently the widths of the series'spectral peaks for the seasonal frequencies are very narrow. It is seen how the "holes" that X11 induces for these frequencies are excessively wide.)

Given that the SA series produces a cyclical signal with too much noise, it would seem that the signal could be improved by removing the noise from the SA series. Thus we replace the decomposition (6.3) by

$$x_t = p_t + s_t + u_t, \tag{6.7}$$

where s_t is as in (6.5) and u_t, for the case of the airline model, is white noise.

From (6.3) and (6.7), $n_t = p_t + u_t$, so that the component p_t is, in fact, the noise-free SA series. This component p_t is often referred to as the trend-cycle component or the short-term trend; it follows, in the four cases, an IMA(2,2) model of the type

$$\nabla^2 p_t = \theta_p(B)a_{pt}, \qquad Var(a_{pt}) = V_p, \tag{6.8}$$

where $\theta_p(B)$ can be factorized as $(1 - \alpha B)(1 + B)$, with the second root reflecting a spectral zero for the frequency π, and α not far from 1 (see the Appendix).

Figure 6.3 plots the SA series, together with the short- and long-term trends (the latter ones obtained with the HP filter). The long-term trend contains little information for short-term analysis, while the SA series simply adds noise to the short-term trend. Using the HP filter on the trend-cycle estimator $\hat{p}_t$, the estimated cycles are displayed in Figure 6.4. Compared to Figure 5.3, use of the trend-cycle instead of the SA series drastically improves the cyclical signal, which becomes much cleaner. Figure 6.5 compares the spectra of the cycles obtained with the two series ($\hat{p}_t$ and $\hat{n}_t$) and, as shown in Figure 6.6, the spectrum of the difference between the cycle spectra based on the trend-cycle and on the SA series is close to that of white noise. So to speak, the band-pass features of the filter implied by the use of the trend-cycle component have improved remarkably; frequencies of no cyclical interest (roughly, those associated with periods of less than two years) are now fully ignored. The improvement of the cyclical signal allows for a clearer comparison of cycles among series, as is evidenced by Figures 6.7 and 6.8. (Considering the scale, Figure 6.4 showed that, for the series AP, the cyclical component has gradually become very small, and hence it has not been included in the last two figures.) In Figure 6.8 it is seen that the series CC, IPI, and CR have similar cyclical patterns, moving roughly in phase.

In summary, we have seen how some important drawbacks of the HP filter can be substantially improved. First, the problem of large revisions, unstable estimation at the end of the series, and poor detection of turning points, are all reduced when the observed series is extended at both ends with forecasts and backcasts computed with adequate ARIMA models. Second, the erraticity of the cyclical signal practically disappears when an AMB trend-cycle estimator is used as input to the filter. We refer to the HP filter that incorporates the two modifications as the modified Hodrick–Prescott (MHP) filter.

In the next chapter, we show how these modifications can be naturally incorporated in a full AMB-type decomposition of the series into a long-term trend, a cycle, a seasonal, and an irregular component. We also show how, in that AMB decomposition, the models for the four components have sensible specifications incorporating model-based features, which ensure compliance with the dynamic structure of the particular series at hand, as well as ad hoc ones that reflect the desirable features of the HP filter. Moreover, the models for the components are such that their aggregate is the parsimonious ARIMA model identified for the observed series, and aggregation of the long-term trend and cycle yields the standard AMB decomposition of Section 3.5. Finally, the MMSE of the cycle in the full model is precisely the MHP filter described above.

Table 6.5 compares the crosscorrelations among the cycles in the three series when the SA series and the trend-cycle are used as inputs. The noise contained in the SA series is seen to reduce the magnitude of the estimated crosscorrelation. The cyclical comovements are better captured with the cycle based on the trend.

Table 6.5. Correlation Among Cycles, Using SA Series (X11) or Trends (SEATS) as Input

Lag	CC-IPI		CC-CR		IPI-CR	
	X11	SEATS	X11	SEATS	X11	SEATS
-4	*	*	*	*	*	*
-3	.20	.36	*	*	*	.23
-2	.31	.55	*	.40	.25	.43
-1	.52	.74	.38	. 58	.42	.61
0	.74	.81	.58	.71	.59	.74
1	.40	.73	.46	.72	.50	.75
2	.31	.57	.44	.67	.46	.69
3	.23	.40	.31	.56	.36	.57
4	*	.25	*	.44	*	.44

(* Not Significant)

One further advantage of using the more stable signal p_t is that it produces a decrease in the size of the revisions in the cyclical estimate for the last periods, as shown in Figure 6.9. The improvement is not spectacular, but it is consistent. Finally, Figure 6.10 displays the 95% confidence interval, based on the revision error, for the cycle estimated for the full period, when the trend-cycle component is used as input.

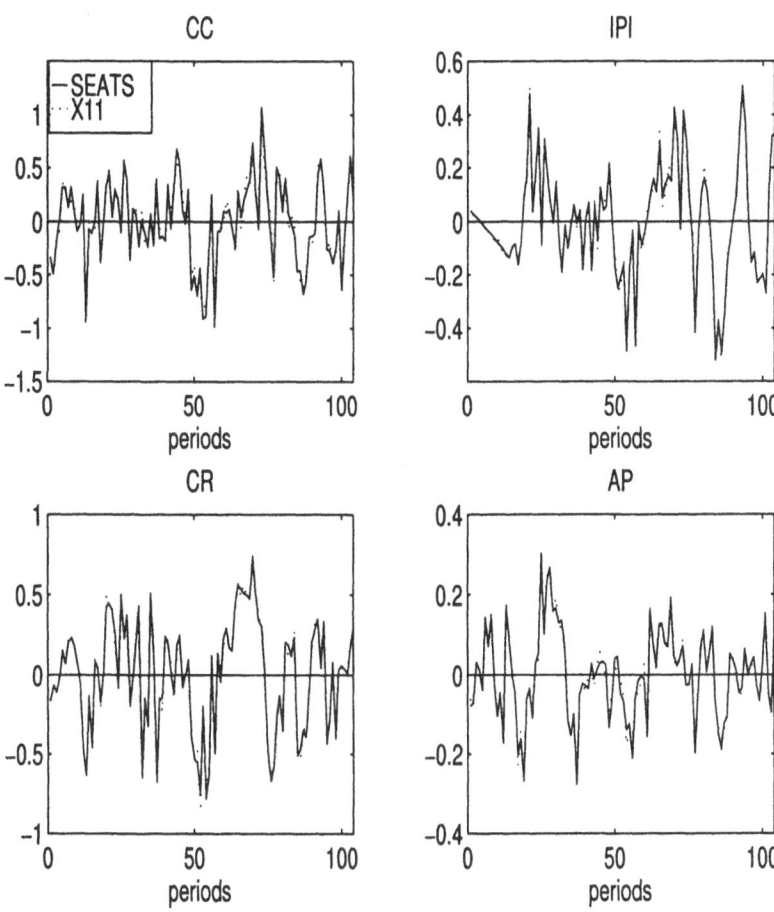

Figure 6.1. HP cycle based on X11 and SEATS-SA series.

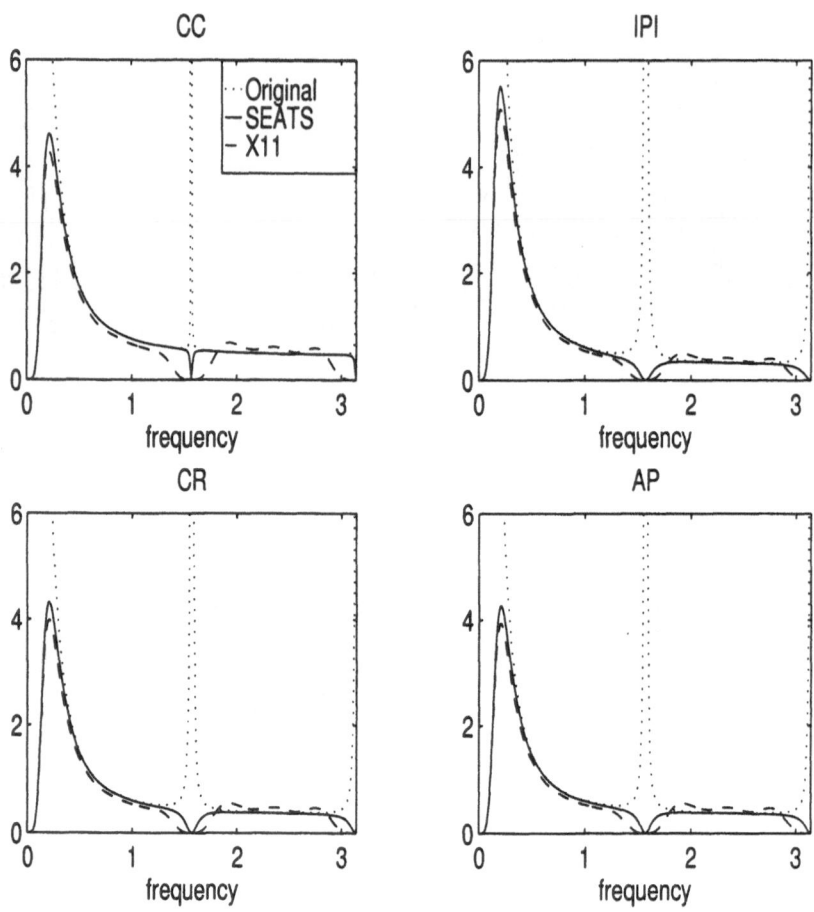

Figure 6.2. Spectrum of cycle based on X11 and SEATS SA series.

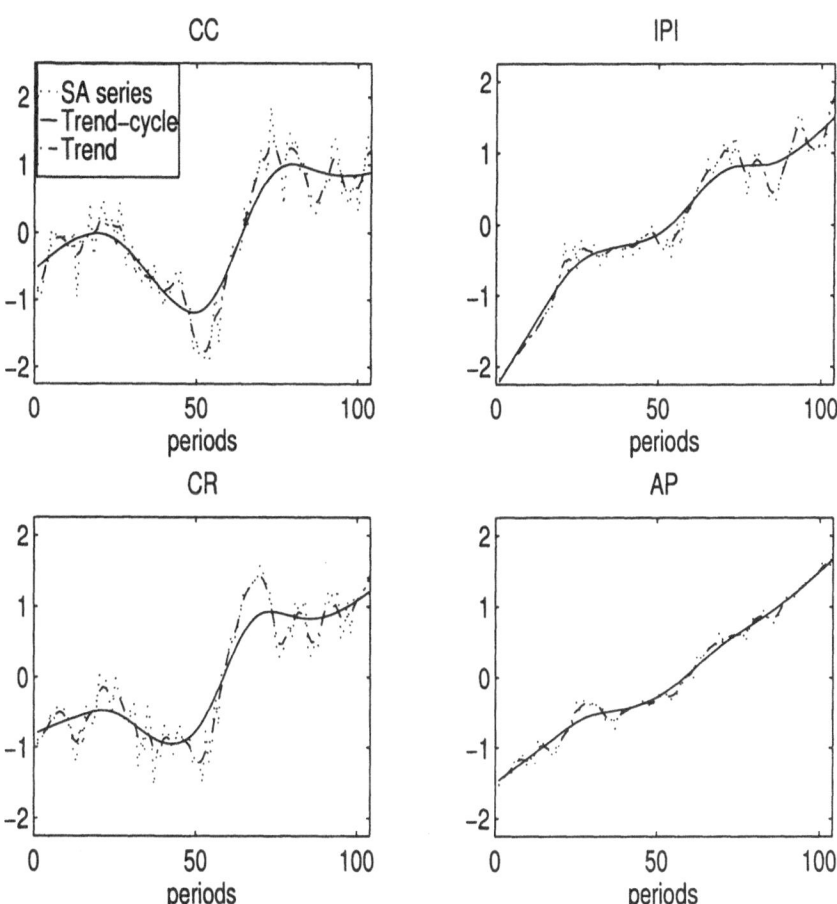

Figure 6.3. Trend and trend-cycle components.

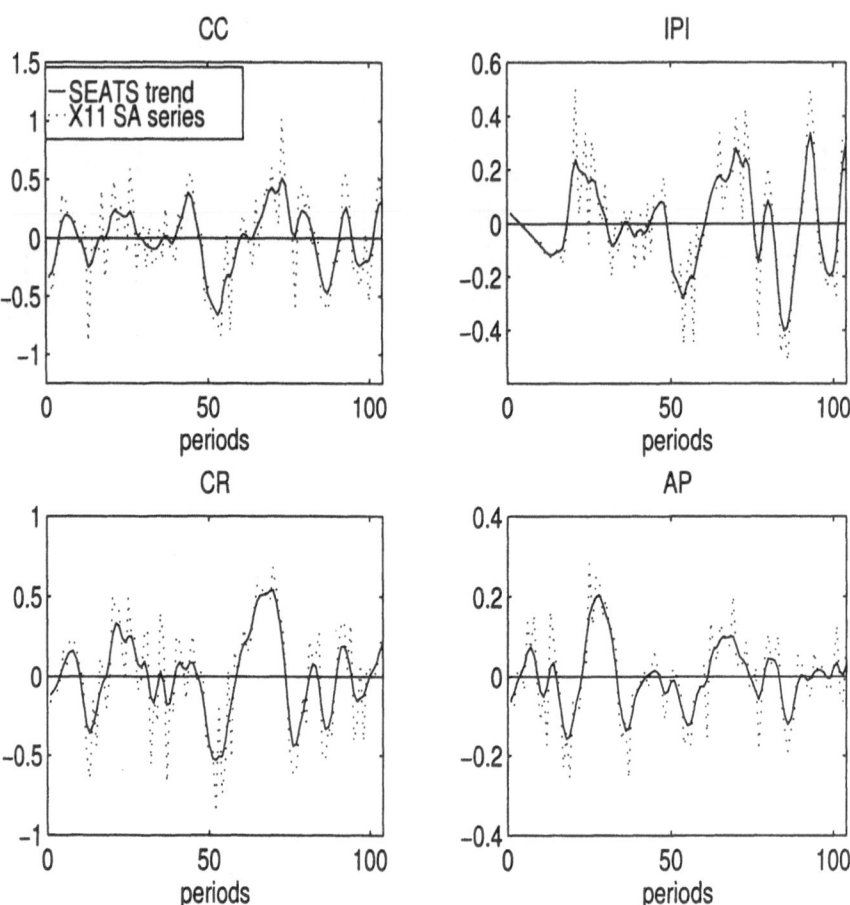

Figure 6.4. HP cycle based on SEATS trend and on X11-SA series.

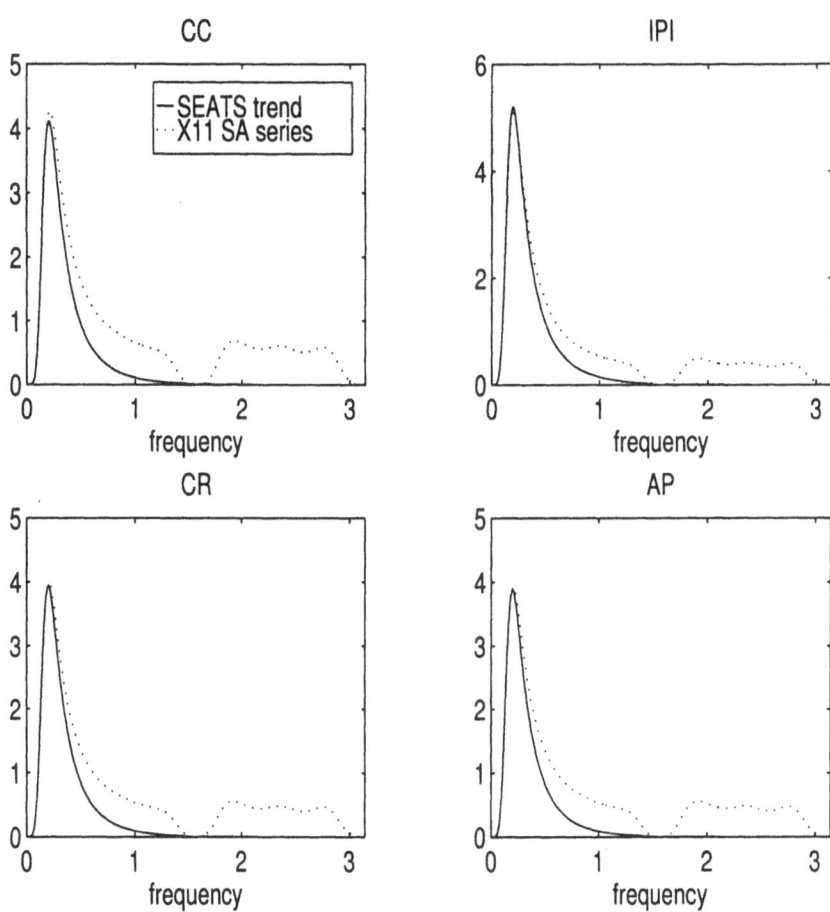

Figure 6.5. Spectrum of cycle (SEATS trend and X11-SA series).

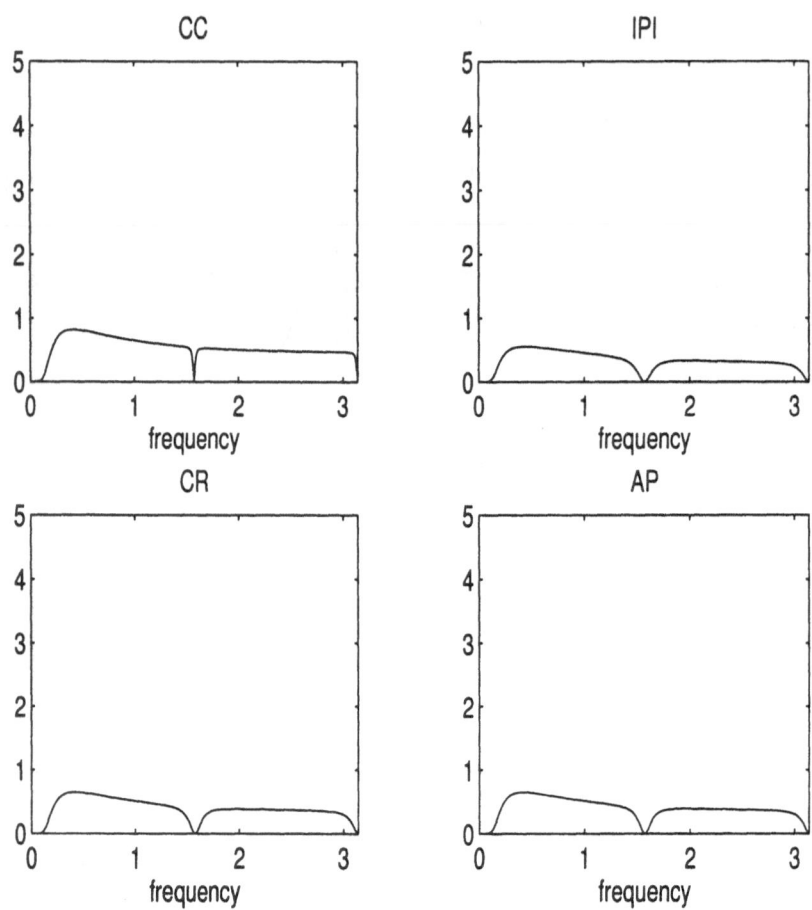

Figure 6.6. Differences between cycles (SEATS trend and X11-SA series).

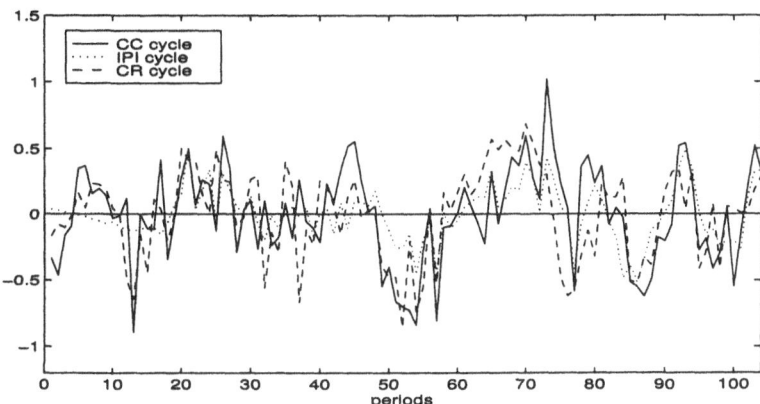

Figure 6.7. HP cycles based on X11-SA series.

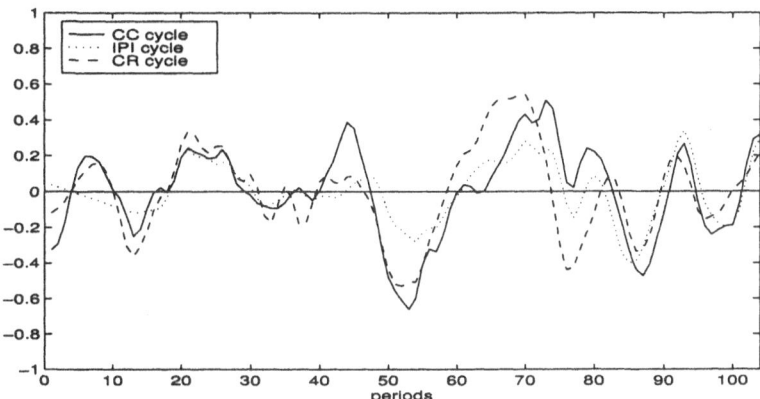

Figure 6.8. HP cycles based on SEATS trend.

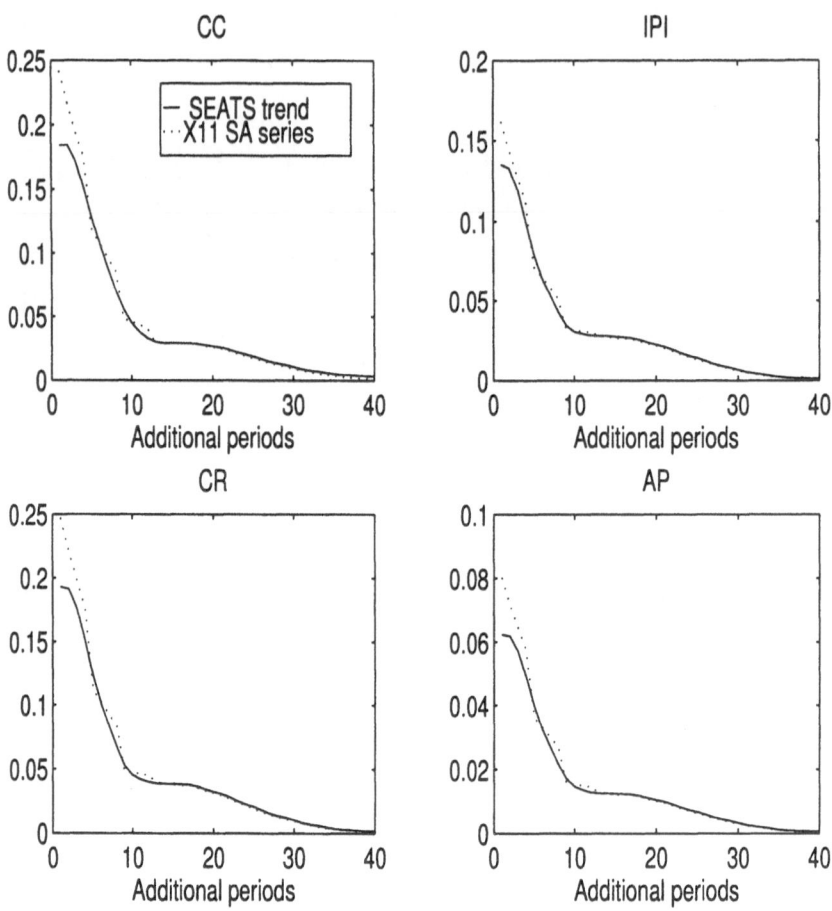

Figure 6.9. Standard deviation of revision from concurrent to final estimation.

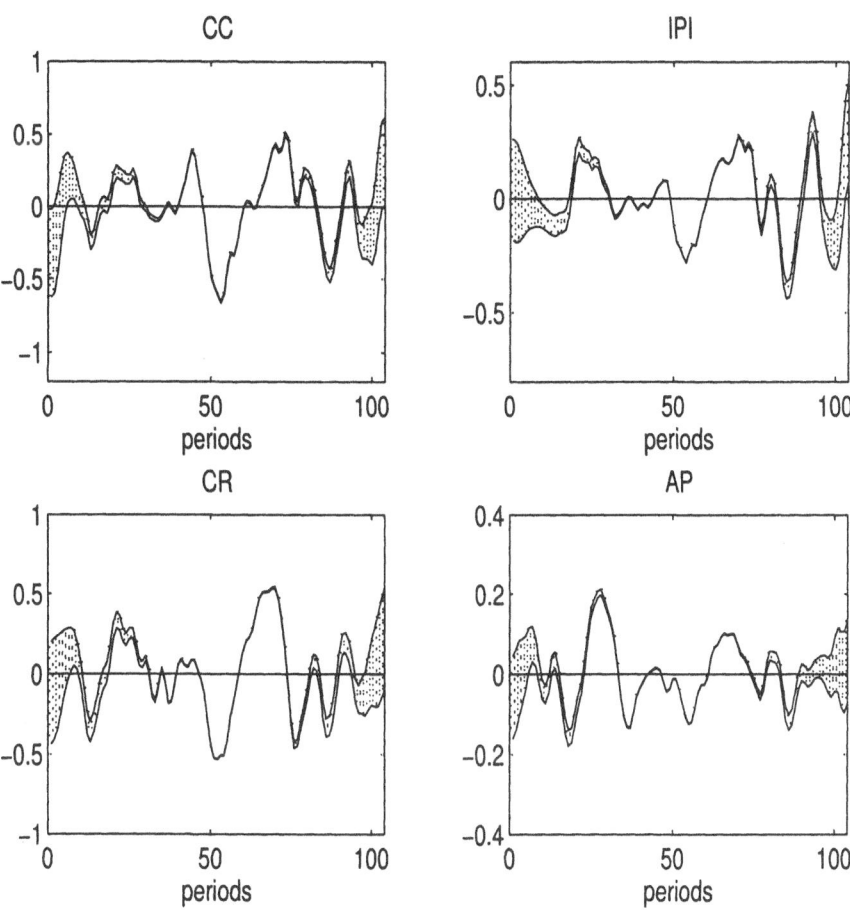

Figure 6.10. 95% Confidence intervals for HP cycle (based on revisions).

7.

Hodrick–Prescott Filtering Within a Model-Based Approach

7.1   A Simple Model-Based Algorithm

What we have suggested in the previous section is to estimate the cycle in steps. First, the AMB method is used to obtain the trend-cycle estimator $\hat{p}_t$ (i.e., the noise-free SA series). In a second step, the HP filter is applied to $\hat{p}_t$.

Assume the observed quarterly series follows the ARIMA model

$$\nabla\nabla_4 x_t = \theta(B)a_t \tag{7.1}$$

with $\theta(B)$ an invertible polynomial. (The discussion extends to other AR structures, but it is greatly simplified using (7.1).) Let (6.8) denote the model for the trend-cycle component obtained from the AMB decomposition (this model is provided in the output of SEATS). From (3.27), the MMSE estimator of p_t is given by the WK filter

$$\hat{p}_t = \left[k_p \frac{\theta_p(B)S}{\theta(B)} \frac{\theta_p(F)\bar{S}}{\theta(F)} \right] x_t, \tag{7.2}$$

where $k_p = V_p/V_a$, and V_p denotes the variance of a_{pt}. The second step consists of obtaining

$$\hat{c}_t = \nu_{HP}^c(B, F)\hat{p}_t, \tag{7.3}$$

where the ν_{HP}^c filter is as in (4.24). Without loss of generality, let us standardize the units of measurement by setting $V_a = 1$, and let $k_c = V_p k_{c(HP)}$, where $k_{c(HP)}$ was defined in (4.24). Then it is obtained that, in terms of the observed series,

$$\hat{c}_t = \left[k_c \frac{\theta_p(B)\nabla\nabla_4}{\theta_{HP}(B)\theta(B)} \frac{\theta_p(F)\bar{\nabla}\bar{\nabla}_4}{\theta_{HP}(F)\theta(F)} \right] x_t. \qquad (7.4)$$

Direct inspection shows that the filter in (7.4) is the AGF of the model

$$\theta_{HP}(B)\theta(B)z_t = \theta_p(B)\nabla\nabla_4 b_t, \qquad (7.5)$$

with $Var(b_t) = k_c$.

It is wellknown (see, e.g., Bell and Hillmer (1984) or Maravall (1987)) that if a series following an ARIMA model is decomposed into signal plus white noise, the MMSE estimator of the noise is given by a filter equal to the autocovariance generating function of the inverse model (multiplied by the variance of the noise). The inverse model was defined in Section 3.1 and is the one that results from interchanging the AR and MA polynomials. Since the model

$$\theta_p(B)\nabla\nabla_4 x_t = \theta(B)\theta_{HP}(B)d_t \qquad (7.6)$$

is the inverse model of (7.5), it follows that $\hat{c}_t$, given by (7.4), is the estimator of the noise in the decomposition of (7.6) into signal plus white noise when the variance of the latter is k_c.

In this way the cycle estimator can be obtained as follows. Let x_t follow the ARIMA model (7.1), and let $\theta_p(B)$ and V_p be the MA polynomial and innovation variance of the model for the trend-cycle p_t in the standard AMB decomposition $x_t = p_t + s_t + u_t$, with s_t and u_t denoting the seasonal and irregular components. To obtain the cycle estimator:

- multiply the AR part of the model for x_t by $\theta_p(B)$; that is,

$$\alpha(B) = \theta_p(B)\nabla\nabla_4;$$

- multiply the MA part of the model for x_t by $\theta_{HP}(B)$; that is,

$$\beta(B) = \theta_{HP}(B)\theta(B).$$

Then the WK filter that yields $\hat{c}_t$ is the estimator of the noise in the decomposition of the model

$$\alpha(B)x_t = \beta(B)a_t \qquad (7.7)$$

into signal plus white noise with the variance of the noise equal to k_c. This filter is directly obtained as the AGF of the model

$$\beta(B)z_t = \alpha(B)a'_t,$$

with $Var(a'_t) = k_c$.

This way of proceeding relies on a white noise assumption for the cyclical component c_t, which is not very appealing. The procedure thus offers a simple algorithm, not a useful model-based interpretation. It is worth noticing that this algorithm will produce estimators for the endpoints different from the ones obtained in the previous procedure (computing first $\hat{p}_t$, and then using the HP filter). This difference is implied by the fact that, in the previous procedure, the forecasts and backcasts, used to extend the series in the Burman–Wilson algorithm described in Section 4.4, are obtained with model (4.20), while in the signal plus noise decomposition of (7.7), they are obtained with model (7.6). In both cases, the models used for forecasting and backcasting x_t are misspecified, since the correct model is (7.1). The difference between the two procedures of course vanishes if the ad hoc forecasts and backcasts are replaced by the appropriate ARIMA ones, that is, if the MHP filter is used instead of the HP one.

7.2   A Complete Model-Based Method; Spuriousness Reconsidered

Looking again at expression (7.4), that provides the cycle estimator for the general model (7.1), another more appealing model-based interpretation is immediately obtained. As seen in Section 3.5, the WK filter only requires the specification of the models for the signal and for the observed series. Comparison with expression (3.26) directly shows that expression (7.4) provides the MMSE estimator of the cycle in model (7.1), when the model for the cycle is of the type

$$\theta_{HP}(B)c_t = \theta_p(B)a_{ct}, \qquad (7.8)$$

with $Var(a_{ct})/Var(a_t) = k_c$. This model-based interpretation of the HP filter, whereby the cycle obtained can be seen as the MMSE estimator of an unobserved component c_t that follows model (7.8), when the observed series follows the general model (7.1), is of some interest. The AR part of the model for the cycle is always the same, and equal to $\theta_{HP}(B)$; it incorporates the "fixed" character of the HP filter. On the other hand, the MA part, equal to $\theta_p(B)$, as well as the variance of the innovation (k_c), will depend on the particular series at hand, and will adapt the filter to the series model.

Therefore, the model for the cycle mixes the band-pass desirable features of the filter with the need to remove seasonality and respect the series'stochastic structure. Model (7.8) is based on the cycle obtained using the trend-cycle component p_t as input. If, instead, the SA series n_t is used, replacing $\theta_p(B)$ and V_p by $\theta_n(B)$ and V_n (the MA polynomial and the innovation variance in the model for the SA series), the interpretation remains the same.

For the four Spanish series, standardized to have $V_a = 1$, Figure 7.1 plots the spectra of the series (dotted line), of its trend-cycle p_t (dashdot line), and of the cycle component c_t (shaded area), when the standard value $\lambda = 1600$ is used. The last component is seen to have well-defined band-pass features that adjust to the width of the spectral peak of the trend-cycle component in the series. Figure 7.2 shows the spectra of the difference between the original series and the cycle spectra (solid line). This difference is clearly made of a long-term trend, a seasonal component, and white noise. The figure also displays the spectra of the original series (dotted line) so that the shaded area represents the series variation captured by the cycle. The decomposition of the series into the two components c_t and $x_t - c_t = m_t + s_t + u_t$, represented in Figure 7.3 (the dotted line is c_t and the continuous line represents $x_t - c_t$), seems perfectly legitimate if interest centers in optimal estimation of the series variation associated with the shaded area of Figure 7.1.

The previous interpretation of the HP filter, applied to the trend-cycle component, as the optimal estimator of the theoretical cyclical component (7.8) when the observed series follows model (7.1), does not specify the models for the rest of the components that have been extracted from the series. The question arises of whether it is possible to give a full model interpretation of the complete decomposition of the series. More specifically, assuming (7.1) is the ARIMA model for the observed series, our two-step previous decomposition can be summarized as follows.

Step I. Decompose the series x_t in the standard AMB manner as $x_t = p_t + s_t + u_t$, where the model for the trend-cycle p_t is of the type (6.8), the model for the seasonal is of the type (6.5), and u_t is white noise. We obtain then the MMSE estimators $\hat{p}_t, \hat{s}_t$ and $\hat{u}_t$.

Step II. Next, the estimator $\hat{p}_t$ is decomposed as in $\hat{p}_t = \hat{m}_t + \hat{c}_t$, where $\hat{m}_t$ is the (long-run) trend estimator, and $\hat{c}_t$ the estimator of the cycle, obtained through the HP filter.

Step II computes estimators directly, without specifying underlying models for the components, and hence the complete decomposition of the series yields $x_t = \hat{m}_t + \hat{c}_t + \hat{s}_t + \hat{u}_t$. Can we rationalize this decomposition as the one obtained from MMSE estimation of orthogonal components in a structural model,

$$x_t = m_t + c_t + s_t + u_t, \tag{7.9}$$

where each component has a sensible model expression, and for which the reduced form (i.e., the model for the aggregate series) is of the type (7.1)? The answer to this question, for a large class of models, is in the affirmative, as we proceed to show.

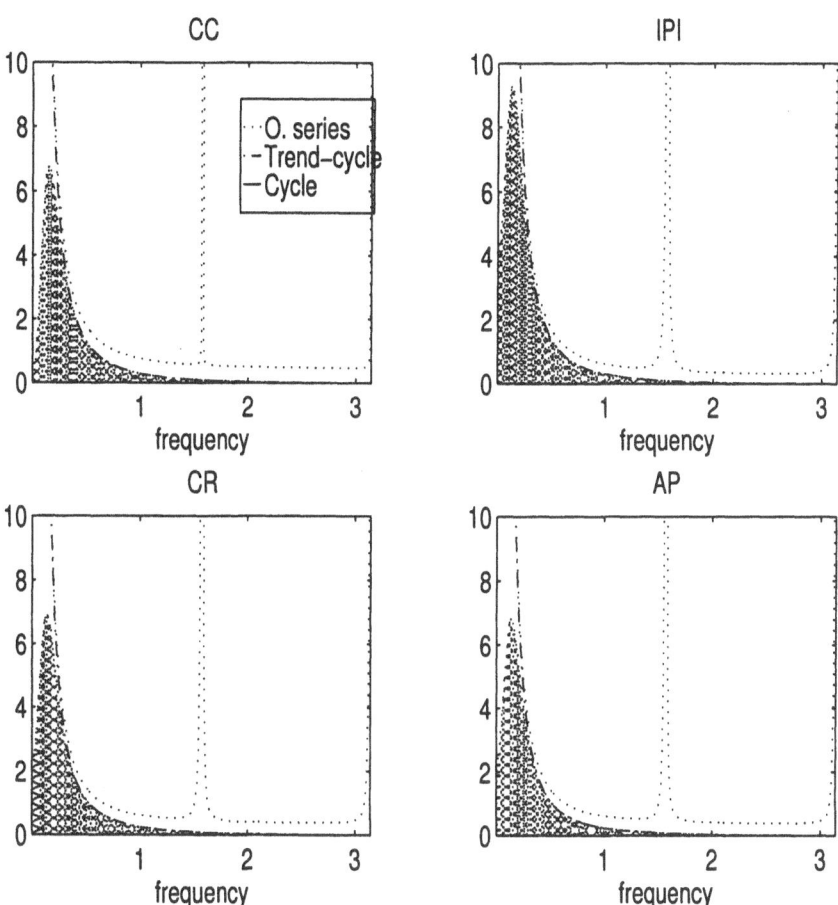

Figure 7.1. Spectra in the model-based interpretation.

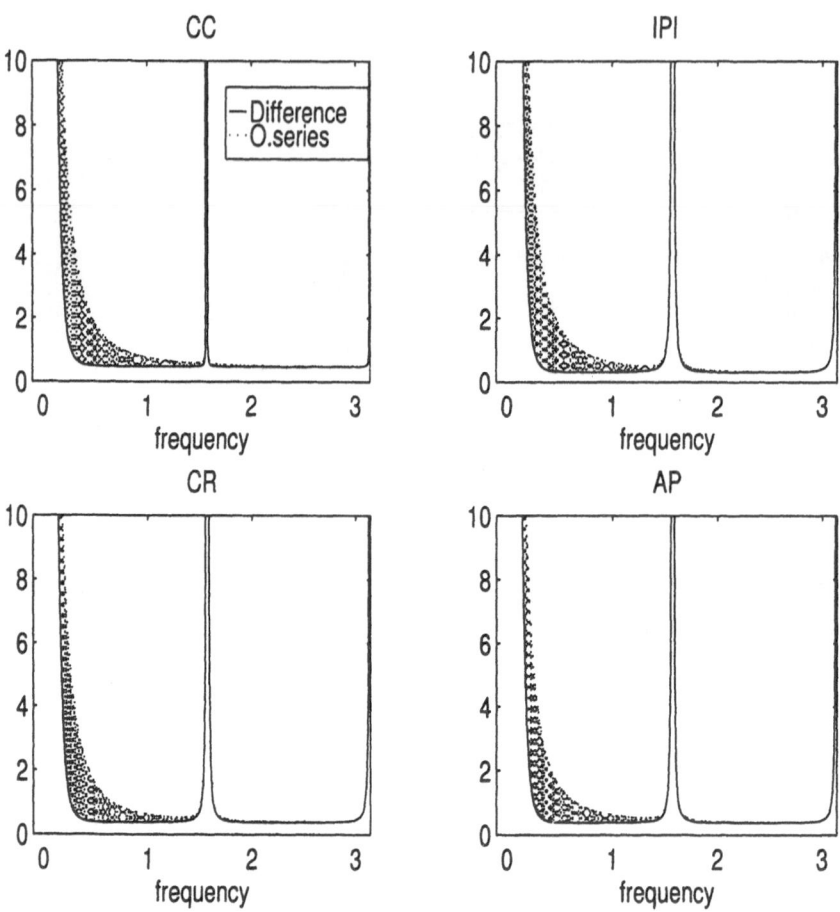

Figure 7.2. Spectra of the difference (original series minus cycle).

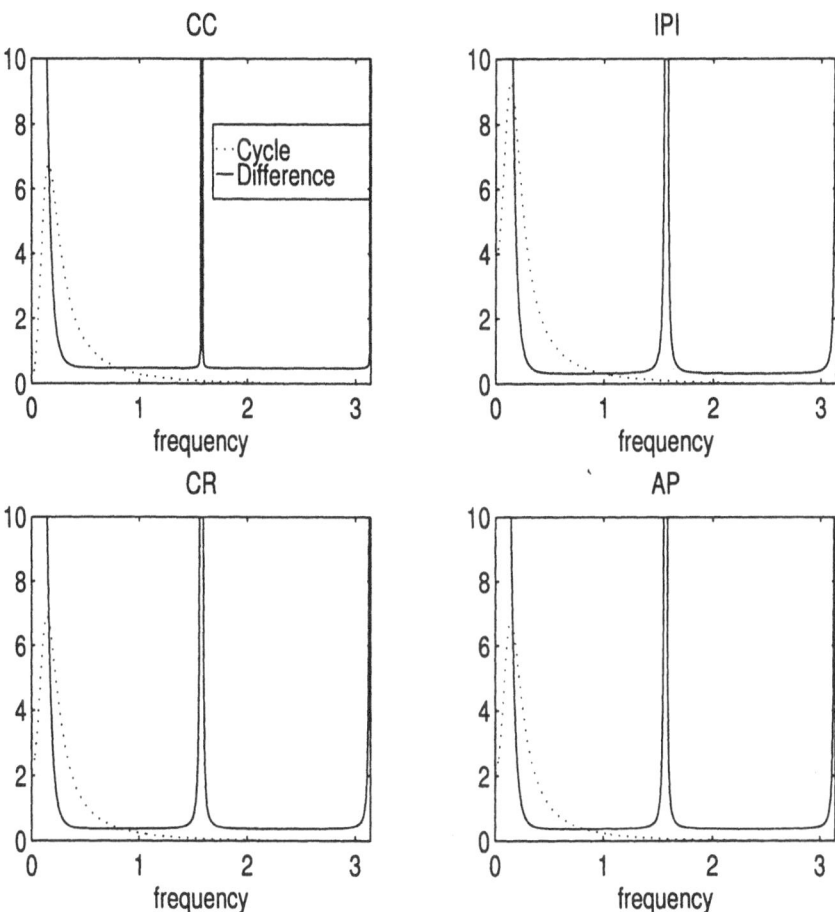

Figure 7.3. Decomposition of the series.

First, if the observed series follows the model (7.1), then the standard AMB decomposition of x_t yields a trend-cycle p_t, a seasonal component s_t, and an irregular component u_t, such that

$$x_t = p_t + s_t + u_t,$$

and the models for the components are of the type

$$\nabla^2 p_t = \theta_p(B)a_{pt}, \qquad Var(a_{pt}) \;=\; V_p, \qquad (7.10)$$
$$Ss_t = \theta_s(B)a_{st}, \qquad Var(a_{st}) \;=\; V_s, \qquad (7.11)$$
$$u_t = \theta_u(B)a_{ut}, \qquad Var(a_{ut}) \;=\; V_u, \qquad (7.12)$$

where $\theta_p(B)a_{pt}, \theta_s(B)a_{st}$, and $\theta_u(B)a_{ut}$ are mutually orthogonal, and normally distributed stationary processes. (If the order of $\theta(B)$ in (7.1) is not larger than 5, then $\theta_u(B) = 1$ and u_t is a white noise irregular. This is indeed the case for the four series considered in the example.) In the second step of the procedure, the HP filter is applied to the MMSE estimator of p_t in the above model, and this yields the estimator of the cycle $\hat{c}_t$, and of the trend $\hat{m}_t$.

Consider now the series x_t that follows model (7.1), and the following unobserved component model,

$$\theta_{HP}(B)\nabla^2 m_t \;=\; \theta_p(B)a_{mt}, \qquad Var(a_{mt}) = V_m, \qquad (7.13)$$
$$\theta_{HP}(B)c_t \;=\; \theta_p(B)a_{ct}, \qquad Var(a_{ct}) = V_c, \qquad (7.14)$$
$$Ss_t \;=\; \theta_s(B)a_{st}, \qquad Var(a_{st}) = V_s, \qquad (7.15)$$
$$u_t \;=\; \theta_s(B)a_{ut}, \qquad Var(a_{ut}) = V_u, \qquad (7.16)$$

where m_t is the trend, c_t the cycle, and s_t and u_t are exactly the seasonal and irregular components of (7.11) and (7.12) (i.e., the ones obtained from the previous AMB decomposition of x_t into p_t, s_t, and u_t). Furthermore, let

$$V_m = V_p k_{m(HP)}, \qquad V_c = V_p k_{c(HP)}, \qquad (7.17)$$

where $k_{m(HP)}$ and $k_{c(HP)}$ were defined in (4.23) and (4.24). The two parameters $k_{m(HP)}$ and $k_{c(HP)}$, as well as the second-order polynomial $\theta_{HP}(B)$, are fixed once the HP parameter λ has been set. (See Section 4.4; for the usual case $\lambda = 1600$, as was seen earlier, standardizing by setting $V_m = 1, V_c = 1600$, it is obtained that $k_{c(HP)} = 1/2001.4, k_{m(HP)} = .7994$, and $\theta_{HP}(B) = 1 - 1.7771B + .7994B^2$.)

From expression (4.21), equating the AGF in both sides of the identity, it is straightforward to obtain the following new identity,

$$\theta_{HP}(B)\theta_{HP}(F) = k_{m(HP)} + \nabla^2\bar{\nabla}^2 k_{c(HP)}. \qquad (7.18)$$

From (4.24), application of the HP filter to obtain the cycle using the trend-cycle estimator $\hat{p}_t$ as input, yields

$$\hat{c}_t = \left[k_{c(HP)} \frac{\nabla^2}{\theta_{HP}(B)} \frac{\bar{\nabla}^2}{\theta_{HP}(F)} \right] \hat{p}_t.$$

Replacing $\hat{p}_t$ by its expression (7.2), and considering (7.17), the estimator $\hat{c}_t$ can be expressed in terms of the observed series as (7.4), which can be rewritten as

$$\hat{c}_t = \left[k_c \frac{\dfrac{\theta_p(B)}{\theta_{HP}(B)}}{\dfrac{\theta(B)}{\nabla\nabla_4}} \frac{\dfrac{\theta_p(F)}{\theta_{HP}(F)}}{\dfrac{\theta(F)}{\bar{\nabla}\bar{\nabla}_4}} \right] x_t,$$

where $k_c = V_c/V_a$. This last expression can be seen as the ratio of two pseudo-AGFs, the one in the denominator being that of the observed series, and the one in the numerator that of the component. As seen in Section 3.4, expression (3.26), this shows that $\hat{c}_t$ is the WK (MMSE) estimator of a cycle c_t that follows the model

$$\theta_{HP}(B)c_t = \theta_p(B)a_{ct}, \qquad Var(a_{ct}) = V_c, \qquad (7.19)$$

when the model for x_t is (7.1).

In an analogous manner, replacing x_t by $\hat{p}_t$ in expression (4.23) and using (7.2), the long-term trend estimator $\hat{m}_t$, obtained by applying the HP trend filter to the trend-cycle estimator $\hat{p}_t$, can be expressed in terms of the observed series x_t as

$$\hat{m}_t = \left[k_m \frac{\theta_p(B)S}{\theta_{HP}(B)\theta(B)} \frac{\theta_p(F)\bar{S}}{\theta_{HP}(F)\theta(F)} \right] x_t$$

$$= \left[k_m \frac{\dfrac{\theta_p(B)}{\theta_{HP}(B)\nabla^2}}{\dfrac{\theta(B)}{\nabla\nabla_4}} \frac{\dfrac{\theta_p(F)}{\theta_{HP}(F)\bar{\nabla}^2}}{\dfrac{\theta(F)}{\bar{\nabla}\bar{\nabla}_4}} \right] x_t,$$

where $k_m = V_m/V_a$. The expression can be seen as a ratio of pseudo-AGFs, which directly shows that $\hat{m}_t$ is the WK (MMSE) estimator of the component m_t, given by the model

$$\theta_{HP}(B)\nabla^2 m_t = \theta_p(B)a_{mt}, \qquad Var(a_{mt}) = V_m, \qquad (7.20)$$

when the model for x_t is (7.1). Noticing that (7.19) and (7.20) are identical to the models for the cycle and trend, given by (7.13) and (7.14), in the full unobserved components model, it follows that the HP estimators of the cycle and trend, when the AMB trend-cycle component is used as input to the filter, are the MMSE of the c_t and m_t components in the full unobserved components model, if the model for the observed series x_t is (7.1).

Now we proceed to show that the sum of the four components,

$$y_t = m_t + c_t + s_t + u_t,$$

equals x_t, and hence follows model (7.1).

Since, by construction, s_t and u_t in the full model (7.13) to (7.16) are the same as those in (7.11) and (7.12), to show that $y_t = x_t$ it suffices to show that $p_t = m_t + c_t$, or equivalently $\nabla^2 p_t = \nabla^2(m_t + c_t)$. From (7.19) and (7.20),

$$z_t = \nabla^2(m_t + c_t) = \frac{\theta_p(B)}{\theta_{HP}(B)} a_{mt} + \frac{\theta_p(B)\nabla^2}{\theta_{HP}(B)} a_{ct}.$$

Since $\nabla^2 p_t$ and z_t are both zero-mean, normally distributed variables, all we need to show is that their autocovariance generating functions are the same. Given the orthogonality of a_{mt} and a_{ct}, the AGF of z_t is equal to

$$\frac{\theta_p(B)\theta_p(F)}{\theta_{HP}(B)\theta_{HP}(F)} V_m + \frac{\theta_p(B)\nabla^2\theta_p(F)\bar{\nabla}^2}{\theta_{HP}(B)\theta_{HP}(F)} V_c$$

$$= \theta_p(B)\theta_p(F)V_p \left[\frac{k_{m(HP)} + \nabla^2\bar{\nabla}^2 k_{c(HP)}}{\theta_{HP}(B)\theta_{HP}(F)} \right],$$

where use has been made of (7.17). In view of (7.18), the term in brackets is 1, so that both $\nabla^2 p_t$ and z_t have the same AGF, equal to $\theta_p(B)\theta_p(F)V_p$, and hence y_t follows model (7.1).

In summary, underlying the two-step procedure we followed in Chapter 6, there is a full unobserved component model whose MMSE estimators yield identical results. It should be mentioned that the equivalence of the two-step and the direct model approach has been shown for historical estimators. For preliminary estimators (i.e., estimators at the endpoints of the series), the direct model approach, implemented via the WK filter (or the Kalman filter), directly offers optimal treatment of endpoints by extending the series x_t (long enough) with the correct model (7.1). Thus the poor behavior of the estimate for recent periods would be improved, and the MMSE estimator in the full model is identical to what we earlier called the MHP filter.

If one uses the seasonally adjusted series, instead of the trend-cycle, as input for the HP filter, the previous unobserved component model is trivially modified. Let n_t denote the seasonally adjusted series

$$n_t = p_t + u_t. \qquad (7.21)$$

From (7.10) and (7.12), the model for n_t is also of the type

$$\nabla^2 n_t = \theta_n(B)a_{nt}, \qquad Var(a_{nt}) = V_n, \qquad (7.22)$$

where $\theta_n(B)$ and V_n are straightforward to obtain from the factorization of

$$\theta_n(B)a_{nt} = \theta_p(B)a_{pt} + \nabla^2\theta_u(B)a_{ut}.$$

Deleting the component u_t, the unobserved component model is now given by (7.15) for the seasonal component, and (7.13) and (7.14), with $\theta_p(B)$ and V_p replaced by $\theta_n(B)$ and V_n, for the cycle and trend models. These replacements are equivalent to adding the noise u_t to the input in the HP filter, which deteriorates the signal, as was seen in Section 6.2.

Thus the sum of the components (7.13) to (7.16), yields the model (7.1) for the observed series, and the MMSE estimators of the components will be identical to the ones obtained with the two-step MHP procedure. Notice, further, that adding m_t and c_t exactly yields the AMB standard decomposition of x_t into trend-cycle, seasonal, and irregular components (the latter two remain, of course, unchanged) and that the AMB trend-cycle component accepts in turn a sensible AMB decomposition into a long-term trend plus (orthogonal) cycle. The two-step procedure thus collapses into direct optimal estimation of components in a fully specified unobserved component model. It is worth pointing out that, in the complete model-based representation, the two components m_t and c_t are canonical (see Section 3.5), because their models contain the MA root $(1 + B)$, present in the polynomial $\theta_p(B)$, and this root implies a spectral zero for $\omega = \pi$ (of course, their sum p_t also presents this feature).

As an example, we show the complete unobserved component model for the car registration (CR) series, one of the four Spanish indicators previously used. The models for the trend-cycle, seasonal, and irregular components in the AMB decomposition of the four series are given in the Appendix. Using the standard value $\lambda = 1600$, and considering (4.22), the full model specification becomes

$$(1 - 1.7771B + .7994B^2)\nabla^2 m_t = (1 + .0662B - .9338B^2)a_{mt},$$
$$(1 - 1.7771B + .7994B^2)c_t = (1 + .0662B - .9338B^2)a_{ct},$$
$$S s_t = (1 + .0383B - .4967B^2 - .4650B^3)a_{st},$$

and u_t white noise, with the component innovation variances given by $V_m = .0386 * 10^{-3}$, $V_c = .0628$, $V_s = .0069$, $V_u = .3695$. It is straightforward to verify that the model for the sum of the components is the airline model

$$\nabla\nabla_4 x_t = (1 - .387B)(1 - .760B^4)a_t,$$

with $V_a = 1$, which coincides with the model identified for CR in Section 5.1.

If one were to use the seasonally adjusted series $\hat{n}_t$ as the input to the HP filter in the 2-step procedure, the unobserved component model interpretation would slightly vary. Replacing V_p by V_n, so that now $V_c = V_n k_{c(HP)}$ and $V_m = V_n k_{m(HP)}$, the unobserved component model is given by

$$x_t = m_t + c_t + s_t,$$

where m_t and c_t are as in (7.13) and (7.14), with $\theta_p(B)$ replaced by $\theta_n(B)$, and s_t is as in (7.15). (The irregular component disappears since it is absorbed by the seasonally adjusted series.) The model for x_t remains unchanged. The effect of these replacements is to add noise to the HP filter input, part of which is passed on to the cyclical signal, so that the cycle obtained from the seasonally adjusted series is in fact equal to the one obtained from the trend-cycle plus some added noise. For the CR series example, as seen in the Appendix, the unobserved component model becomes

$$(1 - 1.7771B + .7994B^2)\nabla^2 m_t = (1 - 1.3215B + .3621B^2)a_{mt},$$
$$(1 - 1.7771B + .7994B^2)c_t = (1 - 1.3215B + .3621B^2)a_{ct},$$

with $V_m = .00041, V_c = .6562$; the model for s_t and the variance V_s remain unchanged, as does the model for x_t. The noise in the input is seen to inflate considerably the variances of the trend and cycle innovations.

For the four series of the application (CC, IPI, CR, and AP), Figures 7.4 to 7.7 display the spectral decomposition into the trend, cycle, seasonal, and irregular components. All of them have sensible shapes, and their sum yields the spectrum of the series, also shown in the figure. Figures 7.8 to 7.11 display the squared gains of the component filters; it is seen how they adapt to the spectral characteristics of each series. The complete specification of the unobserved component model for the four series in the example is given in the Appendix.

An interesting remark concerns the spuriousness question discussed in Section 5.2, which can now be answered in a more precise manner. Insofar as the overall ARIMA model for the observed series fits the data well, it is worth stressing that, because this model and the unobserved components model we have derived from it are observationally equivalent, the latter will also fit the data equally well. The two models, by construction, imply identical joint distribution functions that generate the data (assuming appropriate starting conditions). If the ARIMA model for the observed series is acceptable on empirical grounds, so should be the unobserved component formulation. One may or may not agree, on a priori grounds, with the models specified for the components, although Figures 7.4 to 7.7 show that the trend, cycle, and seasonal components spectra seem quite acceptable, but in no way can the unobserved components model be called spurious.

Be that as it may, our reasoning has been based on a model of the type (7.1), which —as was already mentioned— could be easily generalized by adding an AR polynomial to it. But, given that the proof of the equivalence of the two-step MHP estimator and the MMSE estimator of the cycle in a complete unobserved components model relied on proving that $m_t + c_t$ in the latter was equal to the AMB trend-cycle component p_t, and considering that c_t is stationary, it follows that p_t and m_t have to have the same order of integration for the zero frequency. The trend m_t is I(2), thus p_t also has to be I(2), which implies, in turn, that the overall model for the observed series should be I(2) also (as model (7.1) is). As a consequence, I(1) series, that require only ∇ or ∇_4 differencing to achieve stationarity, would be incompatible with the HP filter. In this sense, the assertion of Harvey (1997) that application of the HP filter to a random walk produces a spurious trend is certainly correct. Yet the issue may be more formal than a serious empirical limitation. To a large extent, the choice of an I(1) or I(2) specification for a particular series depends on the analysts preferences. For example, if a series of innovations a_t is used to generate a random walk model $\nabla x_t = a_t$, it is straightforward to verify that (dropping the first innovation) the series innovations are, in practice, indistinguishable from the ones obtained from the model

$$\nabla^2 x_t = (1 - .99B)a_t, \qquad (7.23)$$

applied to the same series. The data will not be able to discriminate between the I(1) random walk and the I(2) model (7.23). Thus, if an analyst has identified an I(1) model for a series, but wishes to apply the HP filter to estimate the cycle by multiplying by the factors $(1 - B)$ and $(1 - .99B)$, the AR and the MA polynomials, a perfectly sensible model can be specified that would yield an MHP cycle that could not be called spurious.

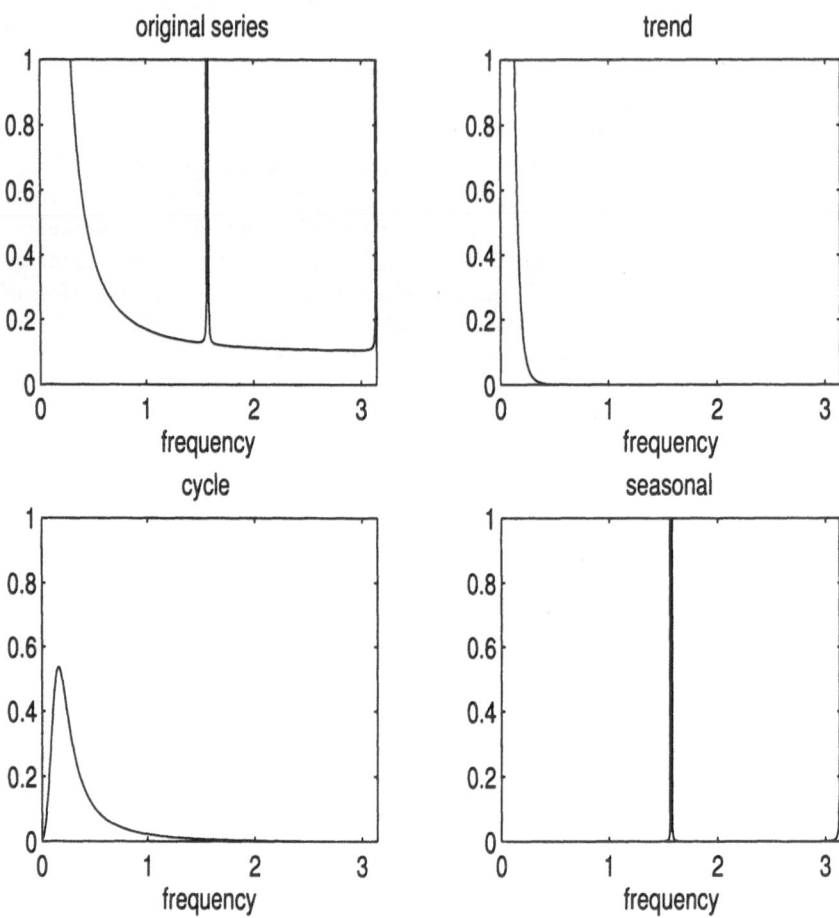

Figure 7.4. Spectra for original series and components: series CC.

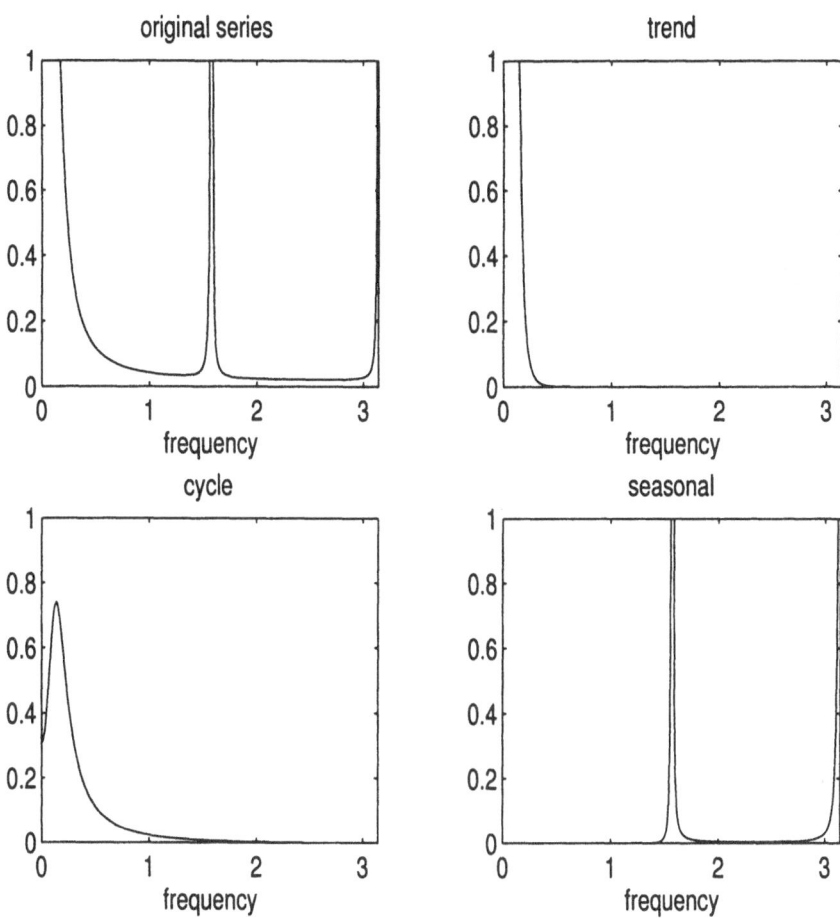

Figure 7.5. Spectra for original series and components: series IPI.

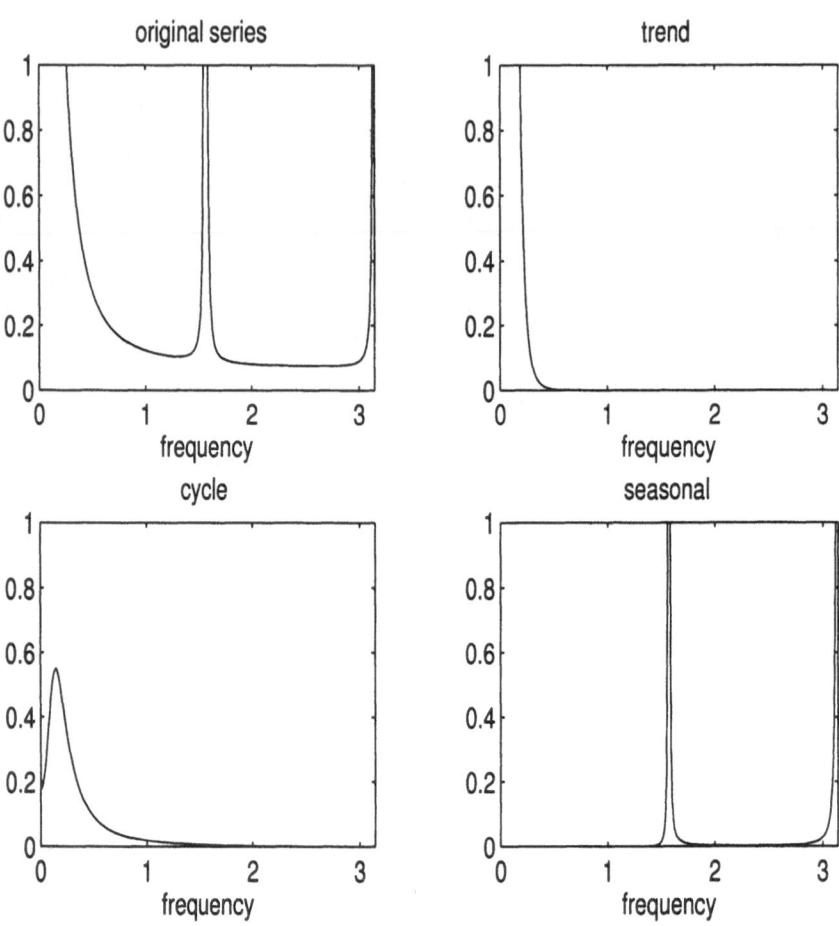

Figure 7.6. Spectra for original series and components: series CR.

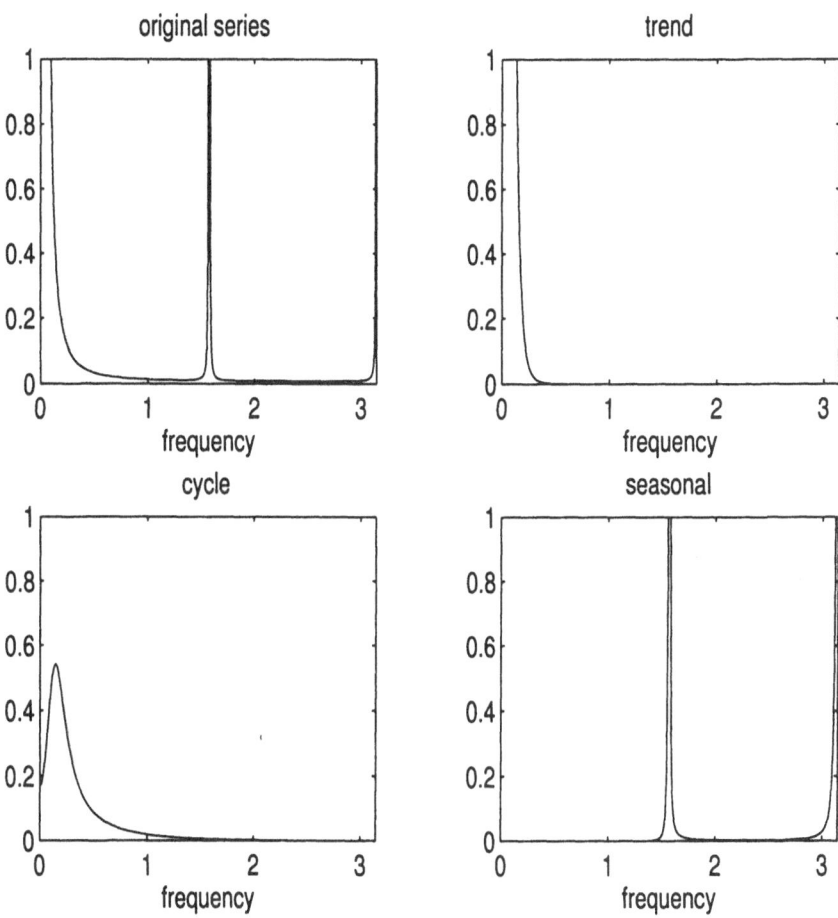

Figure 7.7. Spectra for original series and components: series AP.

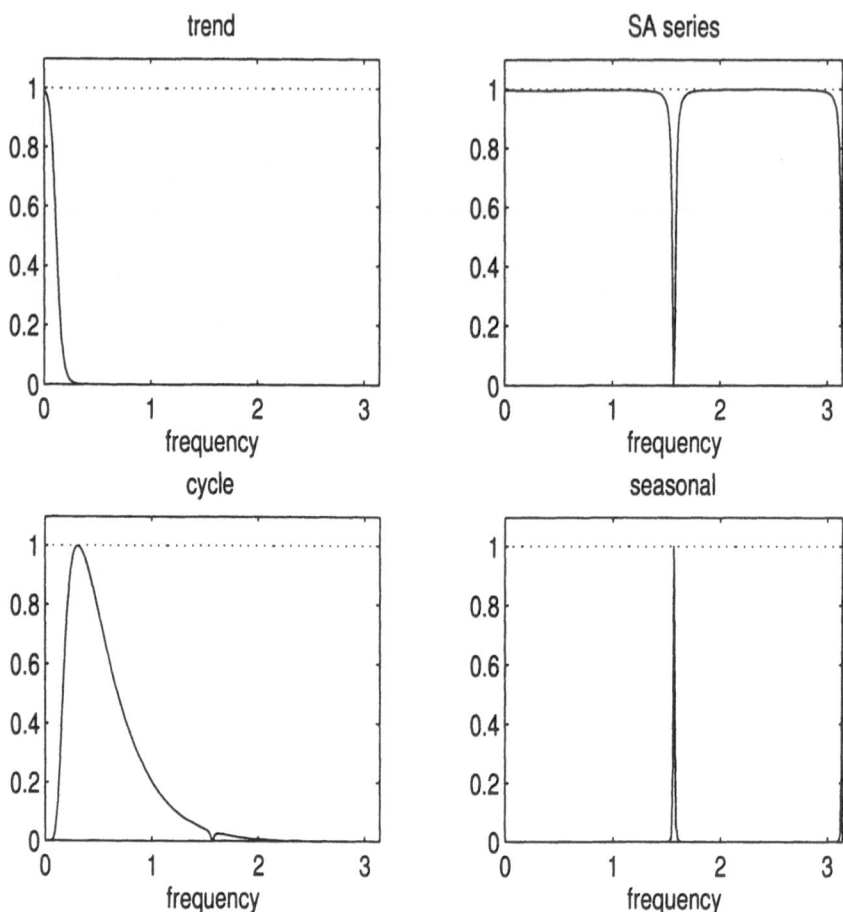

Figure 7.8. Squared gain of filters for components: series CC.

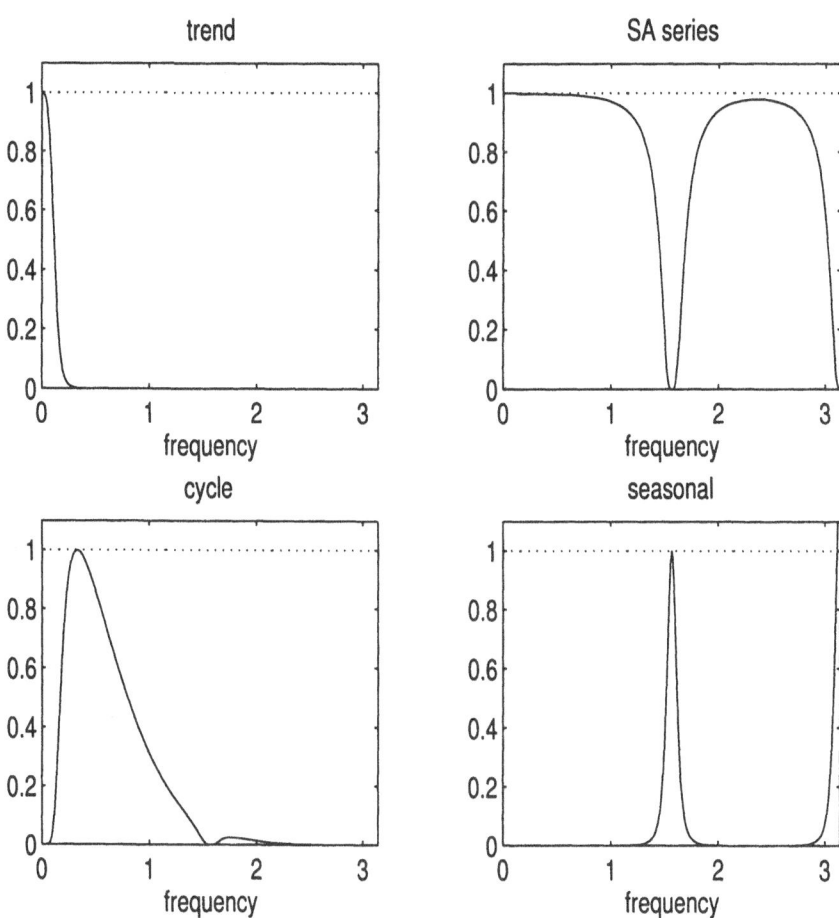

Figure 7.9. Squared gain of filters for components: series IPI.

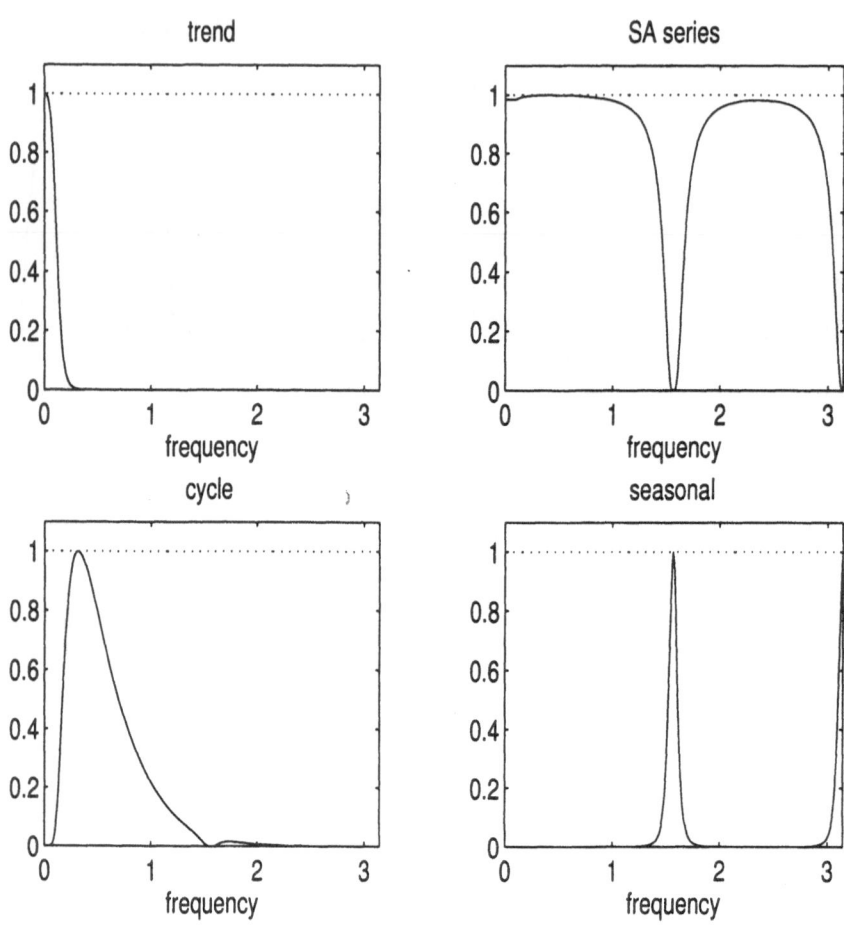

Figure 7.10. Squared gain of filters for components: series CR.

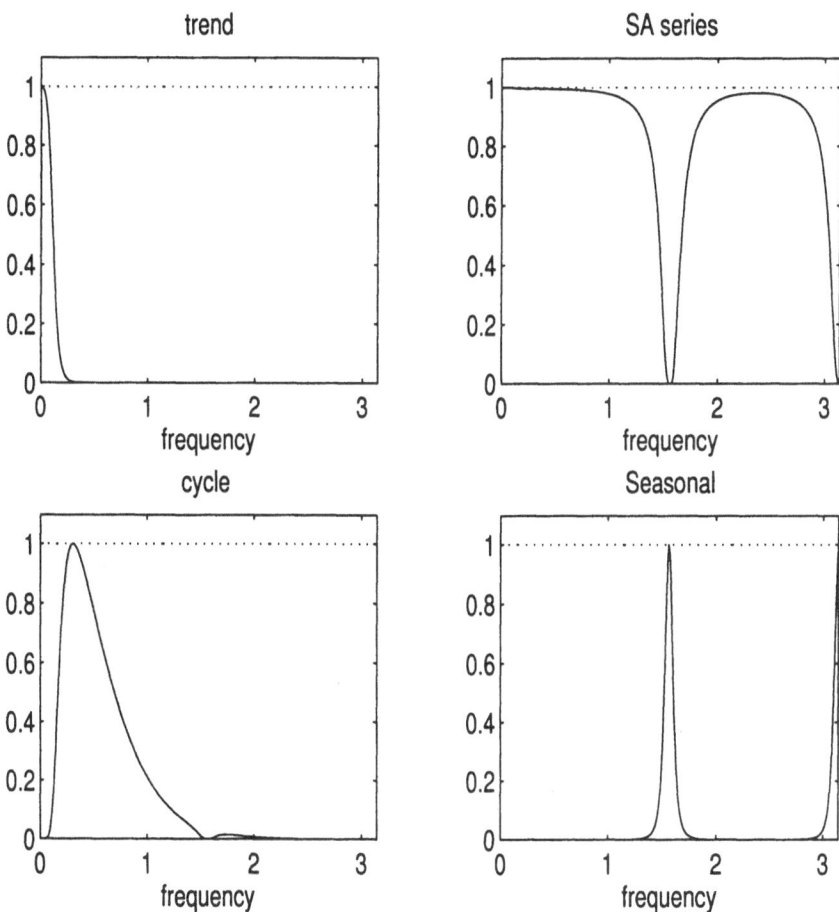

Figure 7.11. Squared gain of filters for components: series AP.

7.3   Some Comments on Model-Based Diagnostics and Inference

One important feature of the model-based procedure is that it automatically overcomes the two limitations of the HP filter mentioned in Chapter 5, namely, the poor performance of the filter at the end of the series (associated with large revisions) and the noisy behavior of the cyclical signal. Revisions are decreased and endpoint estimation is improved, because the series is expanded now with forecasts and backcasts computed with the correct model. On the other hand, the presence of the noninvertible polynomial $\theta_p(B)$ in the MA part of the model for c_t ensures that no noise will contaminate the cycle. For the four series of the example we have been considering, Figures 7.12 to 7.15 compare the standard components, computed with the HP filter applied to the X11 seasonally adjusted series, with the ones obtained as the MMSE estimators in the complete unobserved components model, and serve to illustrate the improvement. (In Section 7.2 we showed that MMSE estimation of c_t in the full unobserved components models is identical to the two-step procedure, whereby the HP filter is applied to the usual AMB trend-cycle estimator, extended at both ends with optimal forecasts and backcasts.) It is seen that the trends obtained with the standard X11-HP and MHP procedures are practically identical, except for the first and final years; this difference at both ends of the series is due to the use of optimal forecasts and backcasts in the model-based procedure. This procedure yields a considerably smoother (much less noisy) cycle than that obtained with the X11-HP procedure. The seasonal components are quite similar, the largest difference occurring for the CC series, for which (as seen in Figure 6.1) X11 clearly overestimates the moving features of the seasonal component. For the X11-HP procedure, the full decomposition yields trend, cyclical, and seasonal components; in the model-based procedure there is an additional component, the irregular, which mainly captures the noise contained in the X11-HP cycle.

From a more general perspective, while (blind) application of the X11-HP filter can be seen as a black-box-type procedure, the model-based approach sets a convenient framework to analyze results, by using well-defined models and estimation criterion (MMSE). The models contain ad hoc features, reflected in the polynomial $\theta_{HP}(B)$, and in $k_{m(HP)}$ and $k_{c(HP)}$ (all three determined from λ, as was seen in Chapter 4), and series-dependent features (such as the polynomial $\theta_p(B)$ and the variance V_p, derived from the overall model for the observed series). Ad hoc features are thus easily incorporated into the AMB approach.

The model-based structure permits us to assess the statistical properties of the cycle such as, for example, its theoretical distribution, as well as the distribution of its optimal estimator. From equation (7.14), the distribution of the theoretical component c_t is easily derived; replacing x_t in expression

(7.4) yields the estimator $\hat{c}_t$ as a function of the innovations in x_t,

$$\hat{c}_t = \left[k_c \frac{\theta_p(B)}{\theta_{HP}(B)} \frac{\theta_p(F)\bar{\nabla}\bar{\nabla}_4}{\theta_{HP}(F)\theta(F)} \right] a_t \tag{7.24}$$

from which the distribution of the MMSE estimator is trivially obtained (it is that of a stationary ARMA process, with the AR polynomial given by the denominator of (7.24), and the MA polynomial by the numerator). Furthermore, since expressions similar to (7.24) can be derived for all component estimators, their joint distribution is easily found. (For nonstationary components, the distributions will be conditional on starting values; see Gómez and Maravall (1993).) The knowledge of these distributions facilitates diagnostics and inference.

First consider diagnostics. Our parametric model fully specifies the distributional features of the processes. Therefore, a natural diagnostic tool is to compare those features (derived from the ARIMA model for the observed series) with the corresponding sample estimates. We choose as illustration an example of some applied concern: the covariance between component estimators.

It is a well-known result (see, e.g., Nerlove et al. (1979)) that, although the theoretical model specifies orthogonal components, the covariance between the components' MMSE estimators will be nonzero. This covariance can be computed as follows.

Equation (7.24) expresses $\hat{c}_t$ as a convergent filter of the innovations a_t. For the trend estimator $\hat{m}_t$, using (4.23) instead of (4.24) with x_t replaced by $\hat{p}_t$, a similar derivation yields

$$\hat{m}_t = \left[k_m \frac{\theta_p(B)}{\theta_{HP}(B)\nabla^2} \frac{\theta_p(F)\bar{S}}{\theta_{HP}(F)\theta(F)} \right] a_t, \tag{7.25}$$

so that the crosscovariance generating function between the two estimators, after simplification, is found to be given by

$$CGF(\hat{m}_t, \hat{c}_t) = k_m k_c \frac{[\theta_p(B)\theta_p(F)]^2 S\bar{S}}{[\theta_{HP}(B)\theta_{HP}(F)]^2 \theta(B)\theta(F)}, \tag{7.26}$$

where, for notational convenience, we have set $V_a = 1$. Direct inspection of (7.26) shows that this CGF is equal to the AGF of the model

$$[\theta_{HP}(B)]^2 \theta(B)z_t = [\theta_p(B)]^2 Sb_t, \qquad Var(b_t) = k_m k_c. \tag{7.27}$$

The model is stationary and hence covariances are bounded. The lag-0 crosscovariance, being equal to the variance of model (7.27), will always be positive.

Be that as it may, insofar as the I(2) component m_t has no proper variance, the theoretical crosscorrelation between the estimators $\hat{m}_t$ and $\hat{c}_t$ is not well defined. For the four series of the example, Table 7.1 displays the sample crosscorrelations among the different components: they are seen to be small with the largest one occurring, as could be expected, between the two components that are stationary, c_t and u_t.

Table 7.1. Crosscorrelations Among Component Estimators: Levels

	CC	IPI	CR	AP
$\hat{m}_t, \hat{c}_t$	.18	−.01	.12	.02
$\hat{m}_t, \hat{s}_t$	.02	−.03	−.02	.00
$\hat{m}_t, \hat{u}_t$	.02	.01	.02	.00
$\hat{c}_t, \hat{s}_t$	−.01	.00	.00	.01
$\hat{c}_t, \hat{u}_t$	.30	.31	.30	.33
$\hat{s}_t, \hat{u}_t$	−.06	.10	.09	.10

Having bounded variances, stationary transformations of the components (namely, $\nabla^2 m_t, c_t, Ss_t$, and u_t) will be crosscorrelated, even asymptotically, in conflict with the theoretical assumptions made for the components. Proceeding as before, the "theoretical" crosscovariance between the MMSE estimators of the different stationary transformations of the components can be derived, and, from that, the crosscorrelation. Therefore, an element for a diagnostic of model adequacy in the model-based procedure can be to compare this theoretical crosscorrelation of the stationary transformations of the MMSE estimators with the ones provided by the sample estimates. The comparison, for the four series of the example and the four components, is given in Table 7.2. The table also contains the standard errors (SE) of the crosscorrelation estimators. These standard errors have been obtained by simulating 1000 series; it is worth mentioning that the model-based framework considerably simplifies simulation because the estimators can be directly generated from the simulated series a_t using the models (7.24), (7.25), and the equivalent ones for the seasonal and irregular components.

Three results seem clear.

- The seasonal component is practically orthogonal to the other components (and hence to the SA series). In fact, the AMB decomposition into trend-cycle, seasonal, and irregular components provides estimators that in practice are close to satisfying the orthogonality assumption made for the theoretical components. Therefore, the conflict between orthogonal components and correlated estimators is more apparent than real and should not be the cause of much concern.

- However, splitting the trend-cycle into trend plus cycle induces negative correlation between the estimators of these two components. Heuristically, this correlation is a reminder of the artificiality of the trend-cycle decomposition. Although the data, summarized in the ARIMA model identified for the series, clearly imply spectral peaks for the zero and seasonal frequencies, which are well captured by the trend-cycle and seasonal components, they have nothing to say about the partition of the zero spectral peak into trend plus cycle.

- The theoretical autocorrelations are in fair agreement with their sample counterparts. In only one case out of 24 the difference (in absolute value) between the estimator theoretical and sample crosscorrelations is larger than 1.96 SE and, even in this case, the associated t-value is moderate. Table 7.2 offers thus a favorable diagnostic concerning model adequacy.

Table 7.2. Crosscorrelations Among Stationary Transformations of Component Estimators: Theretical Estimator and Empirical Estimate

		CC	IPI	CR	AP
$\nabla^2 \hat{m}_t, \hat{c}_t$	Estimator	-.64	-.57	-.59	-.59
	Estimate	-.66	-.65	-.70	-.65
	(SE)	(.09)	(.09)	(.09)	(.09)
$\nabla^2 \hat{m}_t, S\hat{s}_t$	Estimator	-.00	-.00	-.00	-.00
	Estimate	-.00	-.00	-.00	-.00
	(SE)	(.01)	(.01)	(.01)	(.01)
$\nabla^2 \hat{m}_t, \hat{u}_t$	Estimator	-.08	-.05	-.06	-.06
	Estimate	-.07	-.06	-.08	-.08
	(SE)	(.02)	(.02)	(.02)	(.02)
$\hat{c}_t, S\hat{s}_t$	Estimator	.00	.01	.01	.01
	Estimate	.01	.02	-.01	.02
	(SE)	(.02)	(.01)	(.01)	(.01)
$\hat{c}_t, \hat{u}_t$	Estimator	.30	.30	.28	.27
	Estimate	.30	.31	.30	.33
	(SE)	(.04)	(.04)	(.04)	(.04)
$S\hat{s}_t, \hat{u}_t$	Estimator	.05	.13	.13	.13
	Estimate	.09	.14	-.04	.22
	(SE)	(.09)	(.07)	(.07)	(.07)
$\nabla^2 \hat{p}_t, \hat{u}_t$	Estimator	-.16	-.13	-.16	-.17
	Estimate	-.09	-.10	-.16	-.16
	(SE)	(.05)	(.05)	(.05)	(.05)

To see an example of the use of the model in inference, assume we are interested in the following questions.

1. What is the size of the revision in the concurrent estimator and how long should the revision process last in practice?

2. Based only on the size of the revisions, how big —in absolute value— the quarterly growth in the concurrent estimator of the cycle must be in order to reject the hypothesis of zero growth? In other words, when can we accept that the present growth of the cycle is not zero?

Both questions are of applied relevance and can be easily answered by exploiting the model structure. Letting expression (5.3) represent (7.24), the derivation of Section 5.1 can be applied in a straightforward manner. That is, we can write

$$k_c \frac{\theta_p(B)}{\theta_{HP}(B)} \frac{\theta_p(F)\bar{\nabla}\bar{\nabla}_4}{\theta_{HP}(F)\theta(F)} = \xi^-(B) + \xi_0 + \xi^+(F),$$

and, similarly to (5.4), the revision in the concurrent estimator of c_t is equal to

$$r_{t|t} = \xi^+(F)a_{t+1} = \sum_{j=1}^{h} \xi_j a_{t+j},$$

where the second equality relies on the finite truncation. For the four series of the example we have used $h = 250$, more than enough for convergence to the eighth decimal place. The variance of $r_{t|t}$, as well as the periods it takes to have 95% of it removed from the estimator, are straightforward to derive. Specifically, for an integer n, $0 < n < h$, from

$$r_{t|t+n} = \sum_{j=t+n+1}^{h} \xi_j a_{t+j},$$

one can compute the smallest n such that $Var(r_{t|t+n}) \leq .05 Var(r_{t|t})$. Table 7.3 displays, in the first column, the duration in quarters of the revision period measured in the previous way. In all cases the revision period lasts 11 quarters, and hence close to three years. (Notice that in the present case the revision also includes the one associated with the trend-cycle p_t.) The second column of the table shows the standard deviation of the revision in the concurrent estimator expressed as a percentage of the level. For example, for the IPI series, a 67% confidence interval around the concurrent measurement would be on the order of ±1.9 percentage points of the level. The third column of the table presents the standard deviation of the revision as a percentage of the residual standard error; roughly, the size of the revision is about 1/2 that of the one-quarter-ahead forecasts. Altogether, revisions are certainly nonnegligible.

Table 7.3. Size and Duration of the Revision in Concurrent Estimator of Cycle

	Number of Periods to Complete 95% of the Revision	Standard Deviation of the Revision	
		As Percentage of the Level	As percentage of σ_a
CC	11	3.34	44
IPI	11	1.88	58
CR	11	6.46	49
AP	11	2.55	48

As to the question of how big the last quarter growth should be in order to confidently assert it is different from zero, assume that, at time t, the estimator of the cycle for time t is $\hat{c}_{t|t}$. When the next observation becomes available at period $(t + 1)$, the estimator of the cycle for period $(t+1)$ is $\hat{c}_{t+1|t+1}$. Since c_t is measured in logs, differences between (not too distant) periods can be seen as rates of growth. Therefore,

$$\hat{R}_{t+1|t} = \hat{c}_{t+1|t+1} - \hat{c}_{t|t}$$

represents the quarterly rate of growth of the cycle as measured by two consecutive concurrent estimators (for now, we assume that the previous estimator $\hat{c}_{t|t}$ is not updated to $\hat{c}_{t|t+1}$). The final estimator of the growth between these two periods is given by

$$\hat{R}_{t+1} = \hat{c}_{t+1} - \hat{c}_t,$$

so that the error associated with revisions in c_t is equal to:

$$e_{t+1} = (\hat{c}_{t+1} - \hat{c}_{t+1|t+1}) + (\hat{c}_t - \hat{c}_{t|t}).$$

Subtracting from (7.24) its expectation at time t, we can write

$$\hat{c}_t - \hat{c}_{t|t} = \xi^+(F)a_{t+1},$$

and, likewise,

$$\hat{c}_{t+1} - \hat{c}_{t+1|t+1} = \xi^+(F)a_{t+2},$$

so that the error can be expressed as

$$
\begin{aligned}
e_{t+1} &= \xi_1 a_{t+1} + (\xi_1 - \xi_2)a_{t+2} + (\xi_2 - \xi_3)a_{t+3} + \cdots \\
&= \xi_1 a_{t+1} + \sum_{j=1}^{\infty}(\xi_j - \xi_{j+1})a_{t+j+1}. \qquad (7.28)
\end{aligned}
$$

From this expression, the variance of e_{t+1} is easily found, and hence, we can conclude that, using a 90% significance level, only rates of growth of the cycle larger than $1.645\sigma_e$ in absolute value can be assumed to be significantly different from zero.

A more accurate measure of the quarterly growth would be given by $\tilde{R}_{t+1} = \hat{c}_{t+1|t+1} - \hat{c}_{t|t+1}$, where the estimator of the cycle for period t has been revised to take into account the new observation for $(t + 1)$. The previous derivation remains valid, except for the fact that the term $\xi_1 a_{t+1}$ disappears from the rhs of (7.28). For the four series of the example, Table 7.4 presents the borderline growth values (namely, $1.645\, \sigma_e$) below which the (absolute value of the) measured rates of growth cannot be assumed to be significantly different from zero. As before, the values are expressed in percentage points. The table indicates, for example, that the last quarterly growth of the cyclical component in the IPI series (the one estimated with highest precision) needs to be bigger than, approximately, 1%, or smaller than -1%, in order for us to be 90% confident that it cannot be taken as zero.

Table 7.4. Statistical Significance (90% Level) of Quarterly Growth Rate of Cycle (in Percentage Points)

Measured with:	CC	IPI	CR	AP
Concurrent estimators	2.02	1.13	3.76	1.48
Concurrent and revised estimators	2.39	0.98	3.51	1.38

One can easily think of many other extensions. An example could be the computation of optimal forecasts for the components as well as their SE. By construction, these forecasts would be in full agreement with the ones provided by the ARIMA model for the series, directly identified from the data.

Notice that, because the cycle is a zero-mean stationary ARMA(2,2) process with $\theta_{HP}(B)$ as the AR polynomial, its forecasts will gradually approach zero, following damped cosine-type fluctuations in accordance with the roots of $\theta_{HP}(B)$. Added to the limitation implied by the size of the revision error, forecasts of the cycle can be of interest (at most) only for very short horizons.

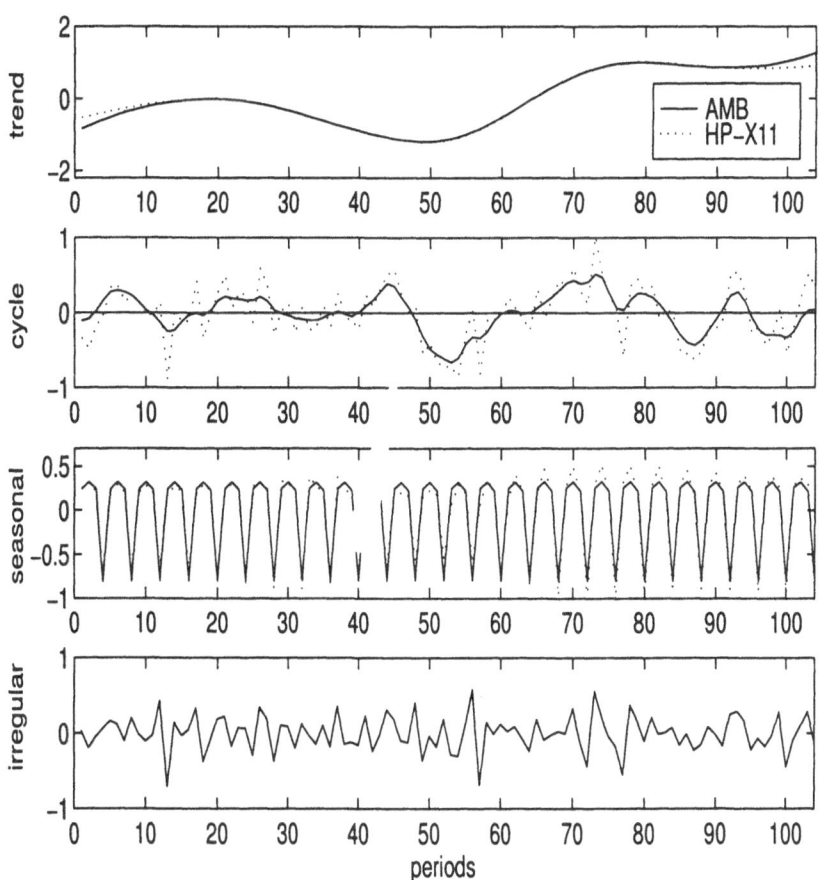

Figure 7.12. AMB and HP-X11: estimated components of CC.

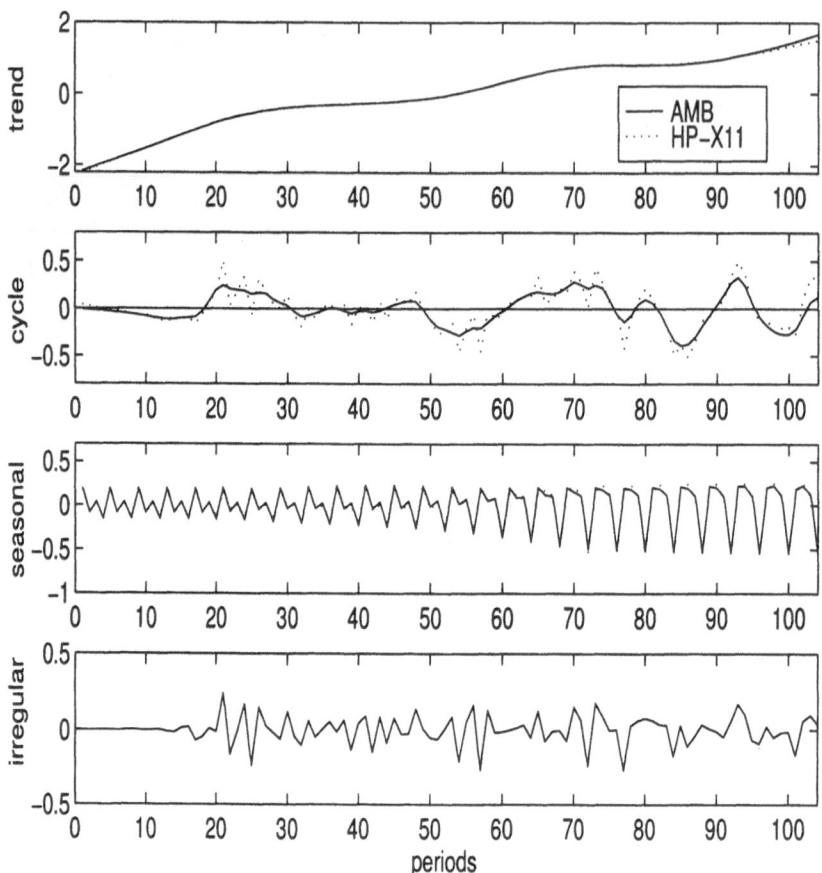

Figure 7.13. AMB and HP-X11: estimated components of IPI.

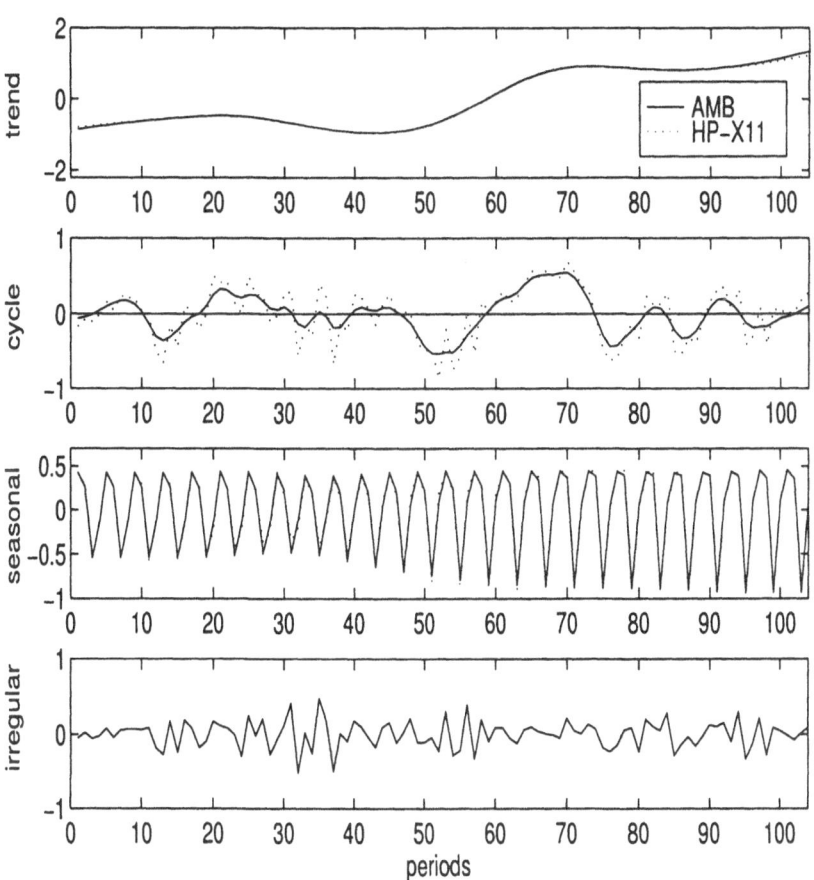

Figure 7.14. AMB and HP-X11: estimated components of CR.

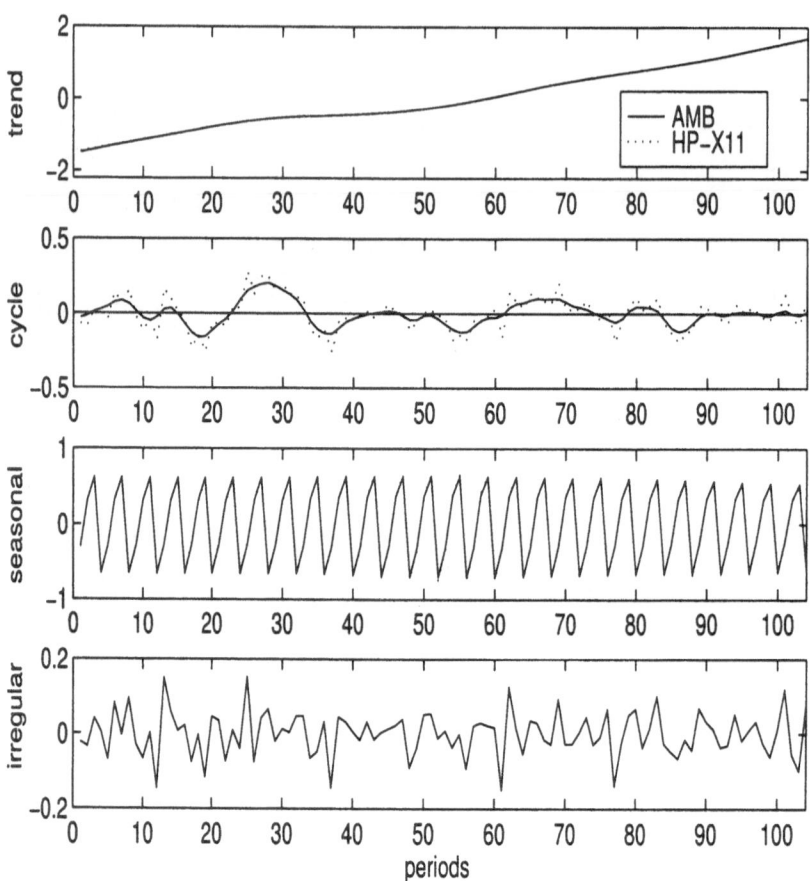

Figure 7.15. AMB and HP-X11: estimated components of AP.

7.4 MMSE Estimation of the Cycle: A Paradox

As before, let the model for the observed series x_t be given by (7.1) and let the trend-cycle component (or noise-free SA series) in the AMB decomposition of x_t be given by (6.8). We have seen that the cycle obtained with the HP filter applied to the trend-cycle component can be seen as the MMSE estimator of a cyclical component c_t that follows model (7.8).

As seen in the appendix, the MA polynomial in (6.8) is always of order 2, and its factorization will be of the type $(1 - \alpha B)(1 + B)$, where α is close to 1 and the root $B = -1$ (a spectral zero for $\omega = \pi$) reflects the "canonical" requirement that the trend-cycle be uncontaminated by noise. The models for the four cycles are seen to be quite similar and, to a rough approximation, the a priori specification $\theta_{HP}(B)c_t = (1 - .9B)(1 + B)a_{ct}$, with $V_c = .07V_a$ can be expected to perform reasonably well. The common structure of the cycle models illustrates how the band-pass approach to filtering can be well accommodated within the model-based approach, so that the advantages offered by a model-based method can be exploited.

Yet the process of computing the MMSE estimator of c_t in the model-based framework presents some conceptual ambiguity, which we proceed to illustrate. If we compute the spectrum of c_t, from model (7.8), for the four Spanish series, and then find the frequency that corresponds to the maximum of the spectrum, the associated period is equal to 10 years. (This period is identical to the one that maximizes the cycle spectrum when the X11-HP filters are applied to a series that follows the airline model.) Therefore, our model would specify a theoretical cycle dominated by the 10-year period.

As we have seen, however, MMSE estimation of a stochastic unobserved component has a distorting effect on the stochastic structure of the component. Equation (7.24) provided the MMSE estimator $\hat{c}_t$ expressed in terms of the innovations a_t in the observed series, from which the spectrum can be easily computed. Denote this spectrum $g_{\hat{c}}(\omega)$. One of the major distortions induced by MMSE estimation, implied by (3.29), is $g_{\hat{c}}(\omega) \leq g_c(\omega)$ for all ω. As a consequence, the MMSE estimator of the cycle will systematically underestimate the stochastic variance of the theoretical cycle.

Figure 7.16 shows the spectra of the trend-cycle and cycle estimators for the four Spanish series. The width of the cycle estimator adjusts now to the width of the spectral peak of the trend-cycle estimator. Comparing Figures 7.1 and 7.16, it is seen that the net effect will be an underestimation of the stochastic variance of the cycle. Perhaps, more disturbingly, MMSE affects the shape of the cycle spectrum and, in particular, the location of its peak. For the four Spanish series, the period associated with the maximum of the cycle estimator spectrum lies between 7.5 and 8 years.

As a consequence, although the theoretical model for the cycle is associated with a main period of 10 years, the (theoretical) MMSE estimator will reduce this period. In other words, if we wish to model a cycle with a

10-year period, our best estimator (in a MMSE sense) will systematically underestimate the period. This creates some ambiguity in terms of deciding which of the two should be taken as the period that characterizes the cycle. For a stubborn analyst wishing that the main period of her model for the cycle be respected by the estimator, MMSE may not be the appropriate criterion. On the other hand, knowing the bias in the underestimation of the period, one could proceed with MMSE estimation, letting the definition of the main cyclical period become a matter of convention.

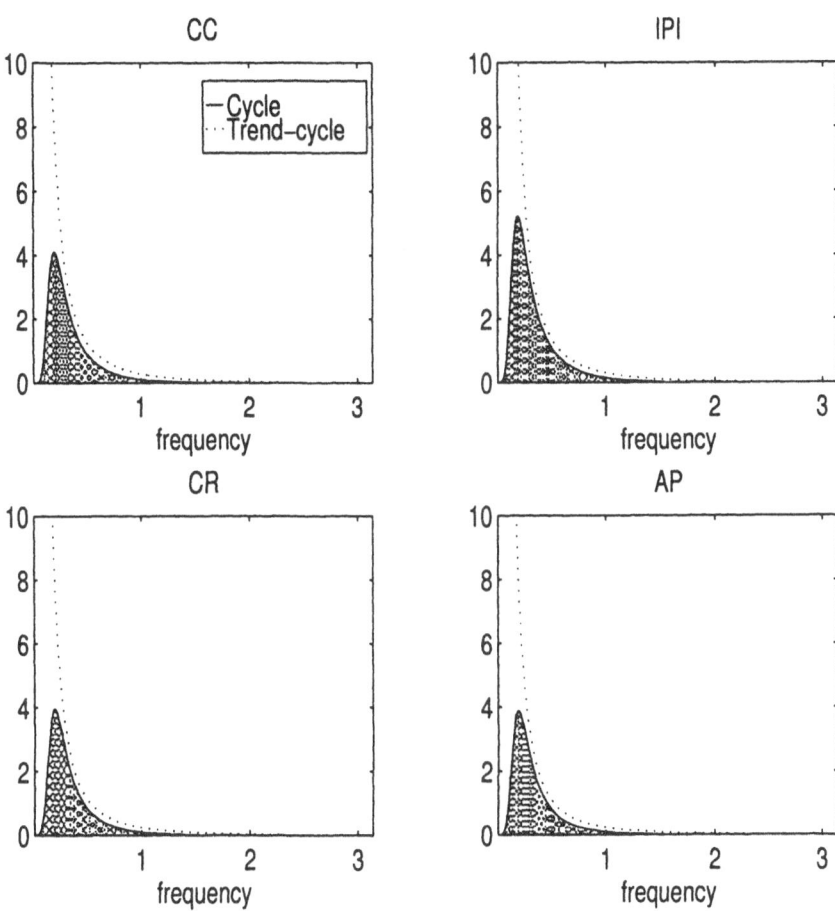

Figure 7.16. Spectra of the trend-cycle and cycle estimators.

Appendix

Complete Unobserved Component Model for the Four Series of the Example

For all cases, the decomposition is given by

$$x_t = n_t + s_t,$$
$$n_t = p_t + u_t,$$
$$p_t = m_t + c_t,$$

where

- x_t = observed series,

- n_t = seasonally adjusted (SA) series,

- s_t = seasonal component,

- p_t = trend-cycle component,

- u_t = irregular component,

- m_t = trend component,

- c_t = cyclical component.

The series are standardized by setting $V_a = 1$; "$w.n.$" denotes a white noise variable.

The components are assumed mutually orthogonal. Given λ (the parameter of the HP filter), all component models are fully derived simply from the ARIMA model for the observed series.

We next list the models for each one of the series and its components. The factorization of the MA polynomials for the trend-cycle and SA series is also given. MHP denotes the modified Hodrick–Prescott filter, as defined in Chapters 6 and 7.

1. Cement Consumption (Series CC)

Model for series:

$$\nabla\nabla_4 x_t = (1 - .405B)(1 - .957B^4)a_t.$$

AMB decomposition into trend-cycle, seasonal, and irregular components

Trend-cycle:

$$\nabla^2 p_t = (1 + .011B - .989B^2)a_{pt} = (1 - .989B)(1 + B)a_{pt},$$

$$V_p = .0856;$$

Seasonal:

$$Ss_t = (1 - .049B - .495B^2 - .455B^3)a_{st}, \quad V_s = .00023;$$

Irregular:

$$u_t = w.n(0, V_u), \quad V_u = .4723;$$

SA series:

$$\nabla^2 n_t = (1 - 1.394B + .401B^2)a_{nt} = (1 - .405B)(1 - .989B)a_{nt},$$

$$V_n = .9675.$$

MHP decomposition into trend plus cycle

Trend:

$$(1 - 1.777B + .799B^2)\nabla^2 m_t = (1 + .011B - .989B^2)a_{mt},$$

$$V_m = .43 * 10^{-4};$$

Cycle:

$$(1 - 1.777B + .799B^2)c_t = (1 + .011B - .989B^2)a_{ct}, \quad V_c = .0685.$$

2. Industrial Production Index (Series IPI)

Model for series:

$$\nabla\nabla_4 x_t = (1 - .299B)(1 - .721B^4)a_t.$$

AMB decomposition into trend-cycle, seasonal, and irregular components

Trend-cycle:

$$\nabla^2 p_t = (1 + .078B - .922B^2)a_{pt} = (1 - .922B)(1 + B)a_{pt},$$

$$V_p = .0975;$$

Seasonal:

$$Ss_t = (1 - .029B - .502B^2 - .527B^3)a_{st}, \quad V_s = .0083;$$

Irregular:

$$u_t = w.n(0, V_u), \quad V_u = .3098;$$

SA series:

$$\nabla^2 n_t = (1 - 1.222B + .277B^2)a_{nt} = (1 - .301B)(1 - .921B)a_{nt},$$

$$V_n = .7932.$$

MHP decomposition into trend plus cycle

Trend:

$$(1 - 1.777B + .799B^2)\nabla^2 m_t = (1 + .078B - .922B^2)a_{mt},$$

$$V_m = .49 * 10^{-4};$$

Cycle:

$$(1 - 1.777B + .799B^2)c_t = (1 + .078B - .922B^2)a_{ct}, \quad V_c = .0779.$$

3. Car Registration (Series CR)

Model for series:

$$\nabla\nabla_4 x_t = (1 - .387B)(1 - .760B^4)a_t.$$

AMB decomposition into trend-cycle, seasonal, and irregular components

Trend-cycle:

$$\nabla^2 p_t = (1 + .066B - .934B^2)a_{pt} = (1 - .934B)(1 + B)a_{pt},$$

$$V_p = .0773;$$

Seasonal:

$$Ss_t = (1 - .038B - .497B^2 - .465B^3)a_{st}, \quad V_s = .0069;$$

Irregular:

$$u_t = w.n(0, V_u), \quad V_u = .369;$$

SA series:

$$\nabla^2 n_t = (1 - 1.322B + .362B^2)a_{nt} = (1 - .388B)(1 - .934B)a_{nt},$$

$$V_n = .821.$$

MHP decomposition into trend plus cycle

Trend:

$$(1 - 1.777B + .799B^2)\nabla^2 m_t = (1 + .066B - .934B^2)a_{mt},$$

$$V_m = .39 * 10^{-4};$$

Cycle:

$$(1 - 1.777B + .799B^2)c_t = (1 + .066B - .934B^2)a_{ct}, \quad V_c = .0618.$$

4. Airline Passengers (Series AP)

Model for series:

$$\nabla\nabla_4 x_t = (1 - .392B)(1 - .762B^4)a_t.$$

AMB decomposition into trend-cycle, seasonal, and irregular components

Trend-cycle:

$$\nabla^2 p_t = (1 + .065B - .935B^2)a_{pt} = (1 - .935B)(1 + B)a_{pt},$$

$$V_p = .0763;$$

Seasonal:

$$S s_t = (1 - .041B - .496B^2 - .463B^3)a_{st}, \quad V_s = .0067;$$

Irregular:

$$u_t = w.n(0, V_u), \quad V_u = .3730;$$

SA series:

$$\nabla^2 n_t = (1 - 1.327B + .367B^2)a_{nt} = (1 - .393B)(1 - .934B)a_{nt},$$

$$V_n = .823.$$

MHP decomposition into trend plus cycle

Trend:

$$(1 - 1.777B + .799B^2)\nabla^2 m_t = (1 + .065B - .935B^2)a_{mt},$$

$$V_m = .38 * 10^{-4};$$

Cycle:

$$(1 - 1.777B + .799B^2)c_t = (1 + .065B - .935B^2)a_{ct}, \quad V_c = .0610.$$

The results can be summarized as follows. All four models for the series are relatively close, in particular for the series CR and AP.

The trend-cycle component is always an IMA(2,2) model, where the MA polynomial can be factorized as

$$\theta_p(B) = (1 - \theta B)(1 + B),$$

with θ close to .95. The model is thus very close to a noninvertible IMA(1,1) model with mean, and with MA root $B = -1$ which implies a monotonically decreasing spectrum with a zero for the frequency $\omega = \pi$. (Incidentally, this model is a particular case of the tangent family of Butterworth filters.)

The seasonal component always follows an ARMA(3,3) model, with the nonstationary AR polynomial $S = 1 + B + B^2 + B^3$. The seasonal innovation variance of the series CC indicates a highly stable seasonal component, while the series IPI contains the most unstable one. All MA polynomials for the seasonal component contain the root $B = 1$ and hence the component spectrum will present a zero for the zero frequency.

The irregular component is always white noise. Between 30 and 50% of the series uncertainty (as measured by the variance of the one-step-ahead forecast error) is caused by the presence of noise. The remaining uncertainty is associated with the stochastic features of the trend-cycle and seasonal components.

The seasonally adjusted series is always an IMA(2,2) model, and the factorization of the MA shows that one of the roots is very close to $B = .95$. As was the case with the trend-cycle, the model is, thus, very close to an IMA(1,1) model with mean. Since the other MA root is moderately small, the model for the SA series is, in the four cases, not far from the popular "random walk plus drift" model (see Stock and Watson (1988)).

The models for the previous components are fully derived from the ARIMA model for the observed series. To further split the trend-cycle into trend plus cycle we need the value of λ (or, equivalently, as was seen in Section 5.2, a prior decision on the main period of the cycle of interest). We use the standard value $\lambda = 1600$, so that the filter will (very broadly) cover cycles with period between 7 and 10 years.

The models for the trend and for the cycle both preserve the MA polynomial $\theta_p(B)$ of the trend-cycle model, and hence will display a spectral zero for the π frequency. The model for the cycle is always a stationary ARMA(2,2) model with the AR polynomial determined by λ. It is also seen that most of the variance of the trend-cycle innovation is absorbed by the cycle itself. Finally, the model for the trend is an ARIMA(2,2,2) process, with the stationary AR polynomial equal to that of the cycle, and a relatively small innovation variance.

References

APEL, M., HANSEN, J., AND LINDBERG, H. (1996), Potential output and output gap, *Quarterly Review of the Bank of Sweden* 3, 24–35.

BANCO DE ESPAÑA (1999), *Boletín Económico*, monthly bulletin, Madrid: Banco de España.

BAXTER, M. AND KING, R.G. (1999), Measuring business cycles: Approximate band-pass filters for economic time series, *The Review of Economics and Statistics* 81(4), 575–593.

BELL, W.R. (1984), Signal extraction for nonstationary time series, *Annals of Statistics* 12, 646–664.

BELL, W.R. AND HILLMER, S.C. (1991), Initializing the Kalman filter for nonstationary time series models, *Journal of Time Series Analysis* 12, 283–300.

BELL, W.R. AND HILLMER, S.C. (1984), Issues involved with the seasonal adjustment of economic time series," *Journal of Business and Economic Statistics* 2, 291–320.

BOLDIN, M.D. (1994), Dating turning points in the business cycle", *Journal of Business*, 67, 97–131.

BOX, G.E.P. AND JENKINS, G.M. (1970), *Time Series Analysis: Forecasting and Control*, San Francisco: Holden-Day.

BOX, G.E.P. AND TIAO, G.C. (1975), Intervention analysis with applications to economic and enviromental problems, *Journal of the American Statistical Association* 70, 71–79.

BOX, G.E.P., HILLMER, S.C., AND TIAO, G.C. (1978), Analysis and modeling of seasonal time series, in Zellner, A. (ed.), *Seasonal Analysis of Economic Time Series*, Washington, DC: US Dept. of Commerce- Bureau of the Census, 309–334.

BROCKWELL,P. AND DAVIS, R. (1987), *Time Series: Theory and Methods*, Berlin: Springer-Verlag.

BRY, G. AND BOSCHAN, C. (1971), *Cyclical Analysis of Time Series: Selected Procedures and Computer Programs*, New York: National Bureau of Economic Research.

BURMAN, J.P. (1980), Seasonal adjustment by signal extraction, *Journal of the Royal Statistical Society A* 143, 321–337.

BURNS, A.F. AND MITCHELL, W.C. (1946), *Measuring Business Cycles*, New York: National Bureau of Economic Research.

BURRIDGE, P. AND WALLIS, K.F. (1984), Unobserved component models for seasonal adjustment filters, *Journal of Business and Economic Statistics* 2, 350–359.

CANOVA, F. (1998), Detrending and business cycle facts, *Journal of Monetary Economics* 41, 475–512.

CHEN, C. AND LIU, L.M. (1993), Joint estimation of model parameters and outlier effects in time series, *Journal of the American Statistical Association* 88, 284–297.

CLEVELAND, W.P. AND TIAO, G.C. (1976), Decomposition of seasonal time series: A model for the X-11 program, *Journal of the American Statistical Association* 71, 581–587.

COGLEY, T. AND NASON, J.M. (1995), Effects of the Hodrick–Prescott filter on trend and difference stationary time series: Implications for business cycle research, *Journal of Economic Dynamics and Control* 19, 253–278.

CRAFTS, N.F.R., LEYBOURNE, S.J. AND MILLS, T.C. (1989), Trends and cycles in British industrial production, 1700-1913, *Journal of the Royal Statistical Society* A, 152, 43–60.

DAGUM, E.B. (1980), The X11 ARIMA seasonal adjustment method. Statistics Canada Catalogue 12-564E.

DANTHINE, J.P. AND GIRARDIN, M. (1989), Business cycles in Switzerland: A comparative study, *European Economic Review* 33, 31–50.

DE JONG, P. (1991), The diffuse Kalman filter", *Annals of Statistics* 19, 1073–1083.

DOLADO, J., SEBASTIÁN, M. AND VALLÉS, J. (1993), Cyclical patterns of the Spanish economy, *Investigaciones Económicas*, XVII (3), 445–473.

ENGLE, R.F. (1978), Estimating structural models of seasonality, in Zellner, A. (ed.), *Seasonal Analysis of Economic Time Series*, Washington, DC: US Dept. of Commerce-Bureau of the Census, 281–297.

EU COMMISSION (1995), The comission services method for cyclical adjustment of government budget balances", *DG II / 401 / 95*-EN.

EUROSTAT (1999), *Monthly Panorama of European Business*, Luxembourg: Official Publications of the European Communities.

FINDLEY, D.F., MONSELL, B.C., BELL, W.R., OTTO, M.C., AND CHEN, B.C. (1998), New capabilities and methods of the X12 ARIMA seasonal adjustment program (with discussion), *Journal of Business and Economic Statistics*, 16, 127–177.

FISCHER, B. AND PLANAS, C. (1998), Large scale fitting of ARIMA models and stylized facts of economic time series, Eurostat Working Paper 9/1998/A/8. To appear in *The Journal of Official Statistics*.

GERSCH, W. AND KITAGAWA, G. (1983), The prediction of time series with trends and seasonalities, *Journal of Business and Economic Statistics* 1, 253–264.

GHYSELS, E. AND PERRON, P. (1993), The effect of seasonal adjustment filters on tests for a unit root, *Journal of Econometrics* 55, 57–98.

GOLDBERG, S. (1967), *Introduction to Difference Equations*, New York: Wiley.

GÓMEZ, V. (1999), Three equivalent methods for filtering nonstationary time series. *Journal of Business and Economic Statistics* 17, 109–116.

GÓMEZ, V. AND MARAVALL, A. (2000a), Automatic modelling methods for univariate series, Working Paper 9808, Research Department, Bank of Spain. To appear as Ch.8 in Peña, D., Tiao, G.C., and Tsay, R.S. (eds.), *A Course in Advanced Time Series Analysis*, New York: Wiley.

GÓMEZ, V. AND MARAVALL, A. (2000b), Seasonal adjustment and signal extraction in economic time series. Working Paper 9809, Research Department, Bank of Spain. To appear as Ch.7 in Peña, D., Tiao, G.C., and Tsay, R.S. (eds.) *A Course in Advanced Time Series Analysis*, New York: Wiley.

GÓMEZ, V. AND MARAVALL, A. (1996), Programs TRAMO and SEATS; Instructions for the user (with some updates), Working Paper 9628, Research Department, Banco de España.

GÓMEZ, V. AND MARAVALL, A. (1994), Estimation, prediction and interpolation for nonstationary series with the Kalman filter, *Journal of the American Statistical Association* 89, 611–624.

GÓMEZ, V. AND MARAVALL, A. (1993), Initializing the Kalman filter with incompletely specified initial conditions in Chen, R. (ed.), *Approximate Kalman Filter (Series on Approximation and Decomposition)*, London: World Scientific 39–62.

GOURIEROUX, C. AND MONFORT, A. (1990), *Séries Temporelles et Modéles Dynamiques*, Paris: Economica.

GRANGER, C.W.J. AND NEWBOLD, P. (1986), *Forecasting Economic Time Series*, New York: Academic.

GRANGER, C.W.J. AND NEWBOLD, P. (1974), Spurious regressions in econometrics, *Journal of Econometrics*, 2, 111–120.

GRENANDER, U. AND ROSENBLATT, M. (1957), *Statistical Analysis of Stationary Time Series*, New York: Wiley.

HABERLER, G. (ed.) (1944), *Readings in Business Cycle Theory*, Philadelphia: Blakiston.

HAMILTON, J.D. (1994), *Time Series Analysis*, Princeton: Princeton University Press.

HAMILTON, J.D. (1989), A new approach to the economic analysis of nonstationary time series and the business cycle, *Econometrica*, 57, 357–384.

HARVEY, A.C. (1997), Trends, cycles and autoregressions, *Economic Journal*, 107, 192–201.

HARVEY, A.C. (1993), *Time Series Models*, Deddington: Philip Allan.

HARVEY, A.C. (1989), *Forecasting Structural Time Series and the Kalman Filter*, Cambridge: Cambridge University Press.

HARVEY, A.C. (1985), Trends and cycles in macroeconomic time series, *Journal of Business and Economic Statistics* 3, 216–227.

HARVEY, A.C. AND JAEGER, A. (1993), Detrending, stylized facts and the business cycle, *Journal of Applied Econometrics* 8, 231–247.

HARVEY, A.C. AND SCOTT, A. (1994), Seasonality in dynamic regression models, *Economic Journal*, 104, 1324–1345.

HARVEY, A.C. AND TODD, P.H.J. (1983), Forecasting economic time series with structural and Box–Jenkins models; A case study, *Journal of Business and Economic Statistics* 1, 299–306.

HATANAKA, M. AND SUZUKI, M. (1967), A theory of the pseudospectrum and its application to nonstationary dynamic econometric models, in Shubik, M. (ed.), *Essays in Mathematical Economics in Honor of Oskar Morgenstern*, Princeton: Princeton University Press.

HENDRY, D.F. (1995), *Dynamic Econometrics*, Oxford: Oxford University Press.

HILLMER, S.C. AND TIAO, G.C. (1982), An ARIMA-model based approach to seasonal adjustment, *Journal of the American Statistical Association* 77, 63–70.

HILLMER, S.C., BELL, W.R. AND TIAO, G.C. (1983), Modelling considerations in the seasonal adjustment of economic time series, in Zellner, A. (ed.), *Applied Time Series Analysis of Economic Data*, Washington, DC: US Department of Commerce-Bureau of the Census, 74–100.

HODRICK, R. AND PRESCOTT, E.C. (1980), Post-war U.S. business cycles: An empirical investigation, mimeo, Pittsburgh: Carnegie-Mellon University.

HYLLEBERG, S., ENGLE, R.F., GRANGER, C.W.J., AND YOO, B.S. (1990), Seasonal integration and cointegration, *Journal of Econometrics* 44, 215–238.

JENKINS, G.M. (1975), The interaction between the muskrat and the mink cycles in North Canada, in *Proceedings of the eighth International Biometric Conference, Editura Acadamiei Republicii Socialiste Romania*, 55–71.

JENKINS, G.M. AND WATTS, D.G. (1968), *Spectral Analysis and Its Applications*, San Francisco: Holden Day.

KAISER, R. AND MARAVALL, A (1999), "Estimation of the Bussines Cycle: a Modified Hodrick-Prescott filter", *Spanish Economic Review* 1, 175-206.

KING, R.G. AND REBELO, S.T. (1993), Low frequency filtering and real business cycles, *Journal of Economic Dynamics and Control* 17, 207–233.

KOHN, R. AND ANSLEY, C.F (1986), Estimation, prediction, and interpolation for ARIMA models with missing data, *Journal of the American Statistical Association* 81, 751–761.

KOOPMAN, S.J., HARVEY, A.C., DOORNIK, J.A., AND SHEPHARD, N. (1996), *Stamp: Structural Time Series Analyser, Modeller and Predictor*, London: Chapman and Hall.

KYDLAND, F.E. AND PRESCOTT, E.C. (1982), Time to build and aggregate fluctuations, *Econometrica* 50, 1345–1370.

LAHIRI, K. AND MOORE, G.H. (eds.) (1991), *Leading Economic Indicators: New Approaches and Forecasting Records*, Cambridge: Cambridge University Press.

MARAVALL, A. (2000), An application of TRAMO and SEATS, Working Paper 9914, Research Department, Bank of Spain. To appear in *Seasonal Adjustment Procedures. Experiences and Perspectives*, Rome: ISTAT.

MARAVALL, A. (1999), Short-Term analysis of macroeconomic time series. in Gerard-Varet A. and Kirman A. (eds.), *Economics: Beyond the Millennium*, Oxford: Oxford University Press.

MARAVALL, A. (1995), Unobserved components in economic time series, in Pesaran, H. and Wickens, M. (eds.), *The Handbook of Applied Econometrics, vol. 1*, Oxford: Basil Blackwell.

MARAVALL, A. (1993), Stochastic linear trends: Models and estimators, *Journal of Econometrics* 56, 5–37.

MARAVALL, A. (1988), A note on minimum mean squared error estimation of signals with unit roots, *Journal of Economic Dynamics and Control* 12, 589–593.

MARAVALL, A. (1987), On minimum mean squared error estimation of the noise in unobserved component models, *Journal of Business and Economic Statistics* 5, 115–120.

MARAVALL, A. AND MATHIS, A. (1994), Encompassing univariate models in multivariate time series: A case study, *Journal of Econometrics*, 61, 197–233.

MARAVALL, A. AND PLANAS, C. (1999), Estimation error and the specification of unobserved component models, *Journal of Econometrics*, 92, 325–353.

MILLS, T.C. (1990), *Time Series Techniques for Economists*, Cambridge: Cambridge University Press.

MORF, M., SIDHU, G.S. AND KAILATH, T. (1974), Some new algorithms for recursive estimation on constant, linear, discrete-time systems, *IEEE Transactions on Automatic Control*, AC 19, 315–323.

NELSON, C.R. AND KANG, H. (1981), Spurious periodicity in inappropriately detrended series, *Econometrica* 49, 741–751.

NELSON, C.R. AND PLOSSER, G.J. (1982), Trends and random walks in macroeconomic time series", *Journal of Monetary Economics* 10, 139–162.

NERLOVE, M., GRETHER, D.M., AND CARVALHO, J.L. (1979), *Analysis of Economic Time Series: A Synthesis*, New York: Academic.

NEWBOLD, P. (1983), Model checking in time series analysis, in Zellner, A. (ed.), *Applied Time Series Analysis of Economic Data*, Washington, DC: US Department of Commerce-Bureau of the Census, 133–143.

OTNES, R.K. AND ENOCHSON, L. (1978), *Applied Time Series Analysis*, New York: Wiley.

PIERCE, D.A. (1979), Signal extraction error in nonstationary time series, *Annals of Statistics* 7, 1303–1320.

PIERCE, D.A. (1978), Seasonal adjustment when both deterministic and stochastic seasonality are present, in Zellner, A. (ed.), *Seasonal Analysis of Economic Time Series*, Washington, DC: US Dept. of Commerce-Bureau of the Census, 242–269.

PLANAS, C. (1997), *Applied Time Series Analysis. Modelling, Forecasting, Unobserved Components, and the Wiener–Kolmogorov Filter*,: Luxembourg: Office for Official Publications of the European Communities.

PRESCOTT, E. (1986), Theory ahead of business cycle measurement, *Carnegie–Rochester Conference Series on Public Policy* 25, 11–66.

SCHUMPETER, J.A. (1939), *Business Cycle*, New York: McGraw-Hill.

SHISKIN, J., YOUNG, A.H., AND MUSGRAVE, J.C. (1967), The X11 variant of the census method II seasonal adjustment program, Technical Paper, 15, Washington, DC.: Bureau of the Census.

SIMS, C.A. (ed.) (1977), *New Methods in Business Cycle Research: Proceedings from a Conference*, Minneapolis: Federal Reserve Bank of Minneapolis.

SINGLETON, K. (1988), Econometric issues in the analysis of equilibrium business cycle models, *Journal of Monetary Economics* 21, 361–386.

STOCK, J.H. AND WATSON, M.W. (eds.) (1993), *New Business Cycles Indicators and Forecasting*, Chicago: Chicago University Press.

STOCK, J.H. AND WATSON, M.W. (1988), Variable trends in economic time series, *Journal of Economic Perspectives* 2, 147–174.

TIAO, G.C. AND TSAY, R.S. (1989), Model specification in multivariate time series, *Journal of the Royal Statistical Society B*, 51, 132–141.

TIAO, G.C. AND TSAY, R.S. (1983), Consistency properties of least squares estimates of autoregressive parameters in ARMA models, *The Annals of Statistics* 11, 856–871.

WHITTLE, P. (1963), *Prediction and Regulation by Linear Least-Squares Methods*, London: English Universities Press.

Author Index

Subject Index

Lecture Notes in Statistics

For information about Volumes 1 to 105, please contact Springer-Verlag

133: Dipak Dey, Peter Müller, and Debajyoti Sinha (Editors), Practical Nonparametric and Semiparametric Bayesian Statistics, xv, 408 pp., 1998.

134: Yu. A. Kutoyants, Statistical Inference For Spatial Poisson Processes, vii, 284 pp., 1998.

135: Christian P. Robert, Discretization and MCMC Convergence Assessment, x, 192 pp., 1998.

136: Gregory C. Reinsel, Raja P. Velu, Multivariate Reduced-Rank Regression, xiii, 272 pp., 1998.

137: V. Seshadri, The Inverse Gaussian Distribution: Statistical Theory and Applications, xi, 360 pp., 1998.

138: Peter Hellekalek, Gerhard Larcher (Editors), Random and Quasi-Random Point Sets, xi, 352 pp., 1998.

139: Roger B. Nelsen, An Introduction to Copulas, xi, 232 pp., 1999.

140: Constantine Gatsonis, Robert E. Kass, Bradley Carlin, Alicia Carriquiry, Andrew Gelman, Isabella Verdinelli, Mike West (Editors), Case Studies in Bayesian Statistics, Volume IV, xvi, 456 pp., 1999.

141: Peter Müller, Brani Vidakovic (Editors), Bayesian Inference in Wavelet Based Models, xi, 394 pp., 1999.

142: György Terdik, Bilinear Stochastic Models and Related Problems of Nonlinear Time Series Analysis: A Frequency Domain Approach, xi, 258 pp., 1999.

143: Russell Barton, Graphical Methods for the Design of Experiments, x, 208 pp., 1999.

144: L. Mark Berliner, Douglas Nychka, and Timothy Hoar (Editors), Case Studies in Statistics and the Atmospheric Sciences, x, 208 pp., 2000.

145: James H. Matis and Thomas R. Kiffe, Stochastic Population Models, viii, 220 pp., 2000.

146: Wim Schoutens, Stochastic Processes and Orthogonal Polynomials, xiv, 163 pp., 2000.

147: Jürgen Franke, Wolfgang Härdle, and Gerhard Stahl, Measuring Risk in Complex Stochastic Systems, xvi, 272 pp., 2000.

148: S.E. Ahmed and Nancy Reid, Empirical Bayes and Likelihood Inference, x, 200 pp., 2000.

149: D. Bosq, Linear Processes in Function Spaces: Theory and Applications, xv, 296 pp., 2000.

150: Tadeusz Caliński and Sanpei Kageyama, Block Designs: A Randomization Approach, Volume I: Analysis, ix, 313 pp., 2000.

151: Håkan Andersson and Tom Britton, Stochastic Epidemic Models and Their Statistical Analysis: ix, 152 pp., 2000.

152: David Ríos Insua and Fabrizio Ruggeri, Robust Bayesian Analysis: xiii, 435 pp., 2000.

153: Parimal Mukhopadhyay, Topics in Survey Sampling, x, 303 pp., 2000.

154: Regina Kaiser and Agustín Maravall, Measuring Business Cycles in Economic Time Series, vi, 190 pp., 2000.

155: Leon Willenborg and Ton de Waal, Elements of Statistical Disclosure Control, xvii, 289 pp., 2000.

156: Gordon Willmot and X. Sheldon Lin, Lundberg Approximations for Compound Distributions with Insurance Applications, xi, 272 pp., 2000.

157: Anne Boomsma, Marijtje A.J. van Duijn, and Tom A.B. Snijders (Editors), Essays on Item Response Theory, xv, 448 pp., 2000.

158: Dominique Ladiray and Benoît Quenneville, Seasonal Adjustment with the X-11 Method, xxii, 220 pp., 2001.